MW00623734

U.S.
COAST GUARD
AIRCRAFT

SINCE 1916

U.S. COAST GUARD AIRCRAFT

SINCE 1916

Arthur Pearcy

Naval Institute Press
Annapolis, Maryland

Dedication: *This book is dedicated to all those intrepid aviators of the US Coast Guard — past and present.*
Semper Paratus

First published in the UK in 1991 by
Airlife Publishing Limited

Published and distributed in the United States of America and Canada by the Naval Institute Press, Annapolis, Maryland 21402

Library of Congress Catalog Card No. 91-62723

ISBN 1-55750-852-6

Printed in England by Livesey Ltd., Shrewsbury.

Contents

ACKNOWLEDGEMENTS

Firstly, my sincere and grateful thanks to all the many US Coast Guard personnel who have kindly hosted Audrey and me during our visits over the years to Coast Guard air stations and other facilities. While attending the gathering of The Ancient Order of Pterodactyls at Elizabeth City USCG air station, North Carolina, in the autumn of 1989, it was a very pleasant surprise to be made an Hon Pterodactyl — No 5 — during an evening I shall never forget. Both Audrey and I were overwhelmed by such a fine gathering of US Coast Guard Aviation past and present. We were privileged to meet and talk to Commander Bruce Melnick USCG, the first Coast Guard astronaut who completed his first Shuttle trip on STS-41 and returned safely.

A great deal of the information and photos illustrating this second USCG aviation volume are in fact an overspill from my earlier volume *A History of US Coast Guard Aviation*. Since that volume Coast Guard Aviation has progressed a step further, has an interesting future to look forward to, and in addition in 1991 will celebrate 75 years of US Coast Guard Aviation.

At US Coast Guard Headquarters in Washington, DC, resides a number of stout and stalwart gentlement who have responded to my many requests for information on the types currently in service. These include Captain David J Connolly (now retired) Chief, Aviation Division, his successor Captain F Power, Commander Dennis C Bosso who filled the vacuum beween Chiefs; Lieutenant Commander Philip A Fallis, HH-60J Facility Manager, and Lieutenant Commander Tom King, HH-65A Facility Manager, Commander John McElwain, and the Chief, Public Affairs, Captain R D Peterson, we have all met during visits to headquarters located on the banks of the Potomac. I nearly forgot my good friend Nicholas G Sandifer, Assistant Chief, Media Relations Branch, Public Affairs at USCG HQ, who was stuck with me through thick and thin during my long research task.

Two US Coast Guard Aviators, now retirées, Captain George Krietemeyer and Captain Marion G Shrode, can hopefully bear a kind mention. They are heavily involved in very hard work with the US National Museum of Naval Aviation and the Pterodactyls respectively. Ptero Les Wiley of Salt Lake City, Utah, very kindly forwarded documentation on Airborne Oil Surveillance System (AOSS) trials while based at Elizabeth City between 1974-78.

My good friend William T Larkins has continued to make available his large collection of memorabilia on US Coast Guard Aviation, and again granted access to his large private collection covering USCG aircraft, both negatives, photographs and colour slides. Peter M Bowers also put his large photo and negative collection at my disposal.

The aircraft manufacturers who still provide the sophisticated aircraft hardware for the Coast Guard were more than cooperative, providing material and photographs as required. They include Lois Lovisolo, Corporate Historian (Grumman Corporation); Susan Miller (Lockheed Aeronautical Systems Company); William Tuttle, Manager, Media Relations (Sikorsky Aircraft); and Harry Gann, Manager, Aircraft Information (Douglas Aircraft Company). To other companies and agencies my grateful thanks.

Photos and colour slides were again obtained from two photo agencies in the United Kingdom: Military Aircraft Photographs (MAP) under the guidance of Brian Pickering, and Aviation Photo News (APN) with Brian Stainer. John Gaffney, Public Affairs Officer USCG at Clearwater Air Station, kindly contributed some excellent aircraft colour slides. Some of John's work can be seen in my previous volume.

Alastair Simpson, Managing Director of Airlife Publishing Limited, has cooperated fully since the birth of this second US Coast Guard Aviation project. With his book editor John Beaton, he has guided, encouraged and shown more than a personal interest. My dear wife Audrey, whose motto is 'If you cannot fight them, join them' has again proof read and counted every word, and has been a tower of strength. My grateful thanks for her continued patience and vigilance.

Lastly, my sincere apologies and thanks to those I have not mentioned, but who contributed to this further tribute to US Coast Guard Aviation, now entering its 75th Anniversary Year with great pride.

INTRODUCTION AND SITREP

'Semper Paratus' — Always Ready

Although tasked by the US Department of Transportation (DoT) in times of peace, the US Coast Guard (USCG) is in fact one of the US five armed forces and has fought in every conflict in which the United States has been involved in since 1790. During 1990 it celebrated its bicentennial, and today it is responsible for the protection of the Maritime Defense Zones (MDZ) which stretch 200 miles out from the US coastline, in addition to a wide variety of peacetime tasks.

Currently over 38,000 strong, with a further 49,000 civilians, reservists and unpaid auxiliaries, the US Coast Guard is a sizeable coastal navy equipped with nearly 250 ships and 2000 boats. It is also the seventh largest naval air arm in the world, operating over 200 fixed and rotary wing aircraft from nearly thirty air stations scattered throughout the mainland United States, Hawaii, Alaska and Puerto Rico. It still maintains thirty-six long range navigation — Loran — units scattered around the world, including some in Europe which are administered by the Commander, US Coast Guard Activities (Europe).

In time of conflict, the US Coast Guard is transferred to the US Navy Department, and has primary responsibility for the Maritime Defense Zones (MDZ) including harbours and navigable waterways. Unarmed helicopters would be tasked with port security and anti-saboteur duties, perhaps deploying with a US Navy fleet, to a cutter force or within state boundaries for the protection of the MDZ. As in previous conflicts, some US Coast Guard pilots and aircrew would be seconded to the US Air Force to administer their skills with combat Search & Rescue (SAR) units.

The US Coast Guard Reserve is very active today with flying operations taking place on Lockheed HC-130H 'Herky Birds' at the air stations located at Sacramento, California; Clearwater, Florida, and Elizabeth City, North Carolina; and include at least one female USCG reserve pilot, who when on duty links up with her husband who is a regular HC-130H jockey. Other females in the regular USCG force fly not only the HC-130H but also helicopters. The Coast Guard Reserve was commissioned in 1987 to equip two reserve squadrons which would support two deployable Hercules squadrons comprised of six aircraft each. During any hostilities they would go to the US Navy flying a projected 920 hours per year.

Current growth in US Coast Guard Aviation in both size and operating tempo, has dictated an increase in aviation manpower.

By the end of Fiscal Year 1990, there will be approximately 860 officer pilots and 3000 enlisted aircrewmen in this, the fifth military arm of the United States. In addition there are eighteen Coast Guard Flight Officers — CGFOs — who man the rear seats of the four Grumman E-2C Hawkeye aircraft. The delivery of two Hawkeyes from the US Customs Service required an increase of four more CGFOs, eight pilots and twenty-one enlisted personnel.

Aviation manpower is maintained within a few percentage points of requirements, being in the 96% area. Nevertheless US Coast Guard Aviation, like many other air arms, is losing many pilots to the commercial airlines. Those with Hercules and Guardian experience are naturally in highest demand. On the credit side many of the new USCG helicopter pilots are ex-US Army, while the odd one has honourably transferred from the US Navy. For some years there has been an exchange of pilots with overseas air arms including the Royal Air Force, Royal Navy and the Canadian Armed Forces (CAF). Currently in the United Kingdom USCG pilots are seconded to a RAF Search & Rescue (SAR) squadron in Scotland; a Royal Navy Search & Rescue (SAR) squadron in Cornwall; a US Air Force Special Operations Squadron (SOS) equipped with the Herky Bird at Woodbridge, Suffolk, while the Deputy Commander at HQ European Activities in London is an aviator. In return both Royal Air Force and Royal Navy helicopter pilots have the opportunity to fly with the US Coast Guard on a reciprocal agreement. The Brits have infiltrated into the Coast Guard over the years and we know a Vice Admiral, now retired, who originated from London and who was an aviator, while a Lieutenant Commander pilot from Southampton is now the USCG Aeronautical Engineering Division (G-EA-2) at headquarters.

In order to maintain an adequate number of pilots, the USCG has a current goal of graduating sixty per year from the US Navy's flight training programme at Pensacola, Florida, and Corpus Christi, Texas. To this is added the transfer of pilots who previously served in other US armed forces. As mentioned earlier, most originate through an attractive direct commission programme for US Army warrant officer helicopter pilots.

The current minimum USCG flight goal for operational duty standby pilots is twenty-five hours per month. However the flight hours tend to be higher for the Lockheed HC-130H pilots than the HH-65A Dolphin pilots. The engine problems of the Dolphin have been put behind to the extent that it is now meeting programme flight hours with a performance that is proving it to be the safest aircraft ever operated in Coast Guard Aviation. All ninety-six HH-65A helicopters have been accepted, the last one on 17 March 1989.

A new air facility located at Charleston, South Carolina, was commissioned and put into operation late 1990, while negotiations with the US Air Force involve assuming the task of

'Hurricane Hunters' with the versatile Herky Bird, to operate from yet a new facility located in Biloxi, Mississippi. The night lab at the USCG Aviation Training Center at Mobile, Alabama, is pouring out pilots proficient in the use of night vision goggles (NVG).

Phasing in of the new Sikorsky HH-60J Jayhawk helicopter as an eventual replacement for the ageing HH-3F Pelican has begun. By the end of September 1990 the USCG had taken delivery of four, with a fifth scheduled in October. During 1991 air stations scheduled to receive the Jayhawk are Elizabeth City, North Carolina, during September, and Traverse City, Michigan, during October. The air stations located at San Francisco, California, and Clearwater, Florida, will receive the type in February and May 1992 respectively. By May 1994 the helicopter should be in service at Cape Cod air station, Massachusetts, and by December at Sitka, Alaska. Kodiak, also in Alaska, is due to take delivery in August 1995, the latter three air station deliveries dependant on the proposed Sikorsky delivery schedule.

Despite heavy demands on aircrew and aircraft, not only in support of the drug war, but with fisheries and patrol activities, search and rescue (SAR) operations, special missions, plus training and aircraft availability, the force is meeting commitments. The huge Exxon tanker disaster in Alaska took manpower and aircraft from USCG air stations throughout the United States.

With nearly 100 fixed-wing aircraft and 150 helicopters listed in the current US Coast Guard Aviation inventory, plus several unmanned airships known as aerostats the future looks fascinating. Several aircraft of each category are undergoing overhaul or modification. The Grumman E-2C Hawkeye AEW aircraft is one of the most capable searchers flown by the US Coast Guard. Fitted with the AWS/APS-125 radar the aircraft are flying nearly 1000 hours per year each. In July 1989, the new E-2C base opened at the Grumman facility located at St Augustine, Florida. Moving the unit, known as CGAW-1 from Norfolk, Virginia, is proving beneficial in more ways than one, and aircraft can easily be deployed to such Caribbean vantage points as Key West, Florida. The 1990 budget carried an appropriation for the conversion and evaluation of an AEW Hercules fitted with an APS-125 rotodome, the E-2C Hawkeye radar, to be fitted to an existing HC-130 '1721'. The Coast Guard sponsored the development of a computer-generated model to determine optimum combination of HC-130H AEW and E-2C aircraft.

A total of six air stations are equipped with the Lockheed HC-130H Hercules in the numerical '1500', '1600' and '1700' series. A total of thirty-one are currently in the USCG inventory. The air stations at Clearwater, Florida and Borinquen, Puerto Rico, operate models equipped with AN/APS-137 FLIR (forward looking infra-red) search radar, as fitted to the US Navy Lockheed P-3C Orion. The AN/APN-215 weather radar is retained. Thirty per cent of HC-130H flying time is spent on search and rescue (SAR) missions. SAMSON — special avionics mission strap-on — developed by Lockheed, is fitted to the port wing. A FLIR turret for night operations has been fitted to Sacramento air station '700' series aircraft which are four of the HC-130H-7 models. It will be used for patrolling fishing areas, law enforcement and other tasks.

Forty-one HU-25A Guardians were procured. Seven have been modified to HU-25B standard configured by AIREYE, being used for oil spill surveillance, ice reconnaissance and law enforcement. All the Guardian aircraft have hard points. Modifications and updates include a computer and wing to carry three surveillance pods — infra-red/ultraviolet (UVLS), line scanners, and SLAR. Budget requests for 1991 were for procurement of the first two of six required activated TV (AGTV) equipped HU-25 aircraft. Nine HU-25C Interceptors are based at Miami air stations located at Opa Locka airport, and at the USCG Aviation Centre at Mobile, Alabama. These are equipped with AN/APG-66 multi-mode radar in the nose as fitted to the General Dynamics F-16 Fighting Falcon, plus a FLIR turret which is mounted in the belly drop hatch. The HU-25A is the basic model of the Guardian.

The Coast Guard procured forty Sikorsky HH-3F Pelican helicopters for Medium Range Recovery Service (MRRS), thirty-six remaining in service by the end of 1990. A total of twelve are based at Clearwater air station, Florida, the largest station now in respect of personnel and aircraft. It also has six HC-130H aircraft. The HH-3F helicopters are heavily involved in the OPBAT Operations — Bahamas, Turks and Caicos mission involved the apprehension of drug traffic. NIGHTSUN is a one million candle power lamp strapped to the Pelican helicopter.

During 1989 the HH-3F fleet was supplemented by the acquisition of six Sikorsky CH-3E helicopters from the US Air Force. These have been overhauled and reworked at Elizabeth City with three operational and on line at Traverse City air station, Michigan. The remainder are retained at Elizabeth City and used for support and spares. They have been modified for SAR missions with an auxiliary fuel tank for increased range, a LORAN 'C' navigation system, and APN-215 radar.

At Opa Locka, Florida, two unique Schweizer RG-8A Condor aircraft are based, these being transferred from the US Department of Defense (DoD) for deployment in connection with the drug war. This 'quiet stealth' type aircraft has a remarkable endurance being more or less a motorised glider, and is virtually noiseless at low altitudes. It is equipped with an avionics pack which includes FLIR and secure radio equipment, plus night vision goggles (NVG). Also based at Opa Locka is a single Spanish-designed CASA 212-300 twin-engined transport, leased to Coast Guard Aviation in full USCG livery for suitability trials in support of helicopter operations from Miami air station. It was delivered mid-1990.

Further north, not too many miles from USCG HQ located on the banks of the Portomac, and based at Washington National airport are two Grumman Gulfstream VIP transports. The flagship is '01', a Gulfstream II delivered in 1969 and used as a Command & Control aircraft by the USCG Commandant, Rear Admiral John W Kime, who replaced Vice Admiral Paul A Yost Jr during May 1990.

As one of the armed forces of the US government, US Coast Guard Aviation in 1991 is celebrating its 75th year of active involvement in the nation's history, its assets adjunct resources to military operational missions, both in peacetime contingency applications and in strategic planning in the defence of America. Despite such rapid progress over recent years, they are not forgetting those intrepid pioneer aviators who went before them, forming the backbone of the current service. One thing that has remained over the seventy-five years are the SOPs — Standing Operation Procedures — some of which have not changed.

When a veteran of US Coast Guard Aviation once reminded the author that 'The rules say we have to go — but there is no rule that says we have to come back', he was repeating a time-honoured Coast Guard adage.

US Coast Guard Aviation Fleet
December 1990

Aérospatiale	HH-65A	Dolphin	96
CASA	C-212-300	Aviocar	1
Dassault-Breguet	HU-25	Guardian	41
Grumman	VC-4A VC-11	Gulfstream	2
Grumman	E-2C	Hawkeye	7
Lockheed	HC-130H	Hercules	31
Sikorsky	HH-3F	Pelican	39
Sikorsky	CH-3H	Sea King	6
Sikorsky	HH-60J	Jayhawk	5
Schweizer	RG-8A	Condor	2
			230

AUTHOR'S NOTE

For some years there has been available a number of excellent detailed volumes with such titles as *United States Military Aircraft Since 1909* (Putnam) and *United States Navy Aircraft Since 1911* (Putnam). Both these volumes are in their second editions which are updated, enlarged and well illustrated. There is a mention in the US Navy volume of US Coast Guard aircraft. During 1990 there appeared the first edition of *US Army Aircraft Since 1947* (Airlife), a welcome companion to the aforementioned two titles. All three are highly recommended.

Since *A History of US Coast Guard Aviation* was published in 1989, being available in time for the 200th anniversary of the US Coast Guard, it was felt that there was a need for *US Coast Guard Aircraft Since 1916*, especially as 1991 will commemorate the 75th anniversary of US Coast Guard Aviation.

This volume, like the three other titles mentioned, and now covering many US armed services aircraft, could over the years ahead be updated and so remain a useful reference to all and sundry. Hopefull it will remain a useful 'book on the shelf' for US Coast Guard aviators past and present, and the Public Affairs Officers on US Coast Guard air stations and other units.

With a breakthrough into more available space for US Coast Guard exhibits at the US National Museum of Naval Aviation located at Pensacola, Florida, the many visitors are being made more aware of the long history of US Coast Guard Aviation. With the help of the Ancient Order of Pterodactyls, more museum exhibits are being added annually and more pioneer USCG aviators added to the Hall of Fame. It is more than coincidental that the museum is located where Elmer F Stone reported for flight training with the US Navy on the first day of April 1916. He qualified as a naval aviator on 22 March 1917, and was eventually designated US Coast Guard Aviator (#1).

There is a feeling in some USCG quarters that the 75th anniversary of US Coast Guard Aviation be celebrated on that day in April when Elmer Stone reported for flight training at Pensacola. As a historian, the author feels that 29 August 1991, 75 years to the day when President Woodrow Wilson, signed into law an act establishing an 'Aerial Coastal Patrol' is the day to celebrate.

The question has been asked many times — 'What was the first Coast Guard aircraft?' During World War I Coast Guard aviators flew US Navy aircraft types, later aircraft were borrowed from the US Navy, but officially the first aircraft owned by the service were the Loening OL-5 and Vought UO-4.

A thought for the future in the fascinating history of US Coast Guard Aviation is 'A Profile of US Coast Guard Aviators'. Many brave men followed in the footsteps of Elmer F Stone, including names such as Donohue (#2); Thrun (#3); Sugden (#4); Parker (#7); Coffin (#8); Eaton (#9), who were not only classmates of Elmer, but who were also awarded the Victory Medal for meritorious war service. While not members of that famous class of 1917, two other outstanding names come to mind and should be mentioned — Wishar (#5) and von Paulsen (#6), who both qualified as USCG aviators in 1920. Other early Coast Guard enlisted men were trained at the US Army aviation school located at Mineola, New York, which brought the small Coast Guard aviation group up to eighteen at the time of the US entry into World War I in 1917.

It seemed an insignificant beginning . . .

Arthur Pearcy Jr AMRAeS
Dakota, 76 High Street
Sharnbrook, Bedfordshire MK44 1PE
England

August 1991

CHAPTER ONE
Seventy-Five Years of Coast Guard Wings

The age of flight was just seven years old when, on 24 August 1916, President Woodrow Wilson signed into law an act establishing an 'Aerial Coastal Patrol'. That was the beginning of US Coast Guard Aviation. Legislation promoted by the Aero Club of America and sanctioned by the US Treasury Department was introduced into the Senate. This proposal provided $1,500,000 for the establishment of the new aerial coast patrol, this to operate as an auxiliary of the Coast Guard. Almost simultaneously, the proposed legislation drafted by Captain McAllister was revamped. When submitted it received congressional approval and was signed into law by the President. This legislation, a part of the US Naval Appropriation Act of 1916, provided for ten Coast Guard air stations to be established along the Atlantic and Pacific coasts, the Great Lakes, and the Gulf of Mexico. However, before any money had been allocated the United States entered World War I and the US Coast Guard became part of the US Navy. During this period several USCG officers and enlisted men underwent training as aviators with the US Navy.

There is no doubt that Coast Guard Aviation owes its beginning to two young officers, Second Lieutenant Norman B Hall and Third Lieutenant Elmer F Stone. Assigned to the USCG cutter *Onondaga* they convinced their commanding officer, Captain B M Chiswell, that what the Coast Guard required was a 'flying surfboat'. After 'selling' the idea fo Coast Guard officials in Washington, DC, Captain Chiswell contacted the pioneer aircraft designer, Glenn H Curtiss, and persuaded him to design a suitable aircraft for rescue work at sea. The result was a triplane flying-boat with a short boat-like hull and with the control surfaces mounted high on the tail booms. This became the forerunner of the famous NC-4.

Coast Guard headquarters took immediate action to incorporate aviation into the service. Captain Charles A McAllister, Chief Engineer, drafted tentative legislation creating an aviation section and US Coast Guard Commandant E P Bertholf queried the US Navy Department concerning the possibility of training

COAST GUARD CUTTER
ONONDAGA

TREASURY DEPARTMENT

UNITED STATES COAST GUARD

Potomac River, April 18, 1916.

My dear Hunniwell:

If practicable, please mail me as soon as convenient plans, specifications and blue prints of a type of motor surfboat which you may regard as best adapted to the following:

Mr. Glenn H. Curtiss, at luncheon with Mr. Newton on the ONONDAGA last Sunday, suggested that it might be practicable to convert a surfboat into a flying boat with wings and motor so arranged that they could be quickly eliminated when the boat lighted on the water and within a few minutes it would be, instead of a flying boat, an ordinary motor surfboat. If the lifeboat is better adapted, send lifeboat. He promised to think about it and I am going to try to encourage him.

If it is possible to perfect something of that kind I believe it would be the biggest find for the Coast Guard of the century and might be the means of saving hundreds of lives. Maybe if you could hear them say nonchalantly that they are now building machines capable of lifting 20 tons, you would not be quite so skeptical as I ~~know~~ believe you to be at present.

Sincerely,

[illegible]

THIS LETTER CONTAINS THE GERM FROM WHICH COAST GUARD AVIATION SPRUNG.

This copy of the original letter written by Captain Chiswell to construction officer Hunniwell at Coast Guard Headquarters, was discovered by the author in the archives held at USCG HQ in Washington DC. It contains the first specification for a 'Flying Life Boat' for use by the Coast Guard. *(USCG)*

USCG officers as pilots. The US Navy agreed initially to accept two Coast Guard officers for training at the newly-established Naval Air Station at Pensacola, Florida. On the first day of April

1916, Second Lieutenant Charles E Sugden and Third Lieutenant Elmer F Stone received orders to report to that station.

Lieutenant Norman B Hall, who with Captain Benjamin M Chiswell and Lieutenant E F Stone was responsible for the introduction of the aviation idea, by virtue of his education and qualifications as a professional naval architect, was ordered on 28 October 1916, to the Curtiss Aeroplane & Motor Co factory at Hammondsport, New York, to study aircraft engineering. The company also had a flying school located at Harbor Point, Newport News, close to where the cutter *Onondaga* had been based.

The year 1916, although important for the birth of Coast Guard Aviation, was an ominous one for the United States and indeed for the free world. On the continent of Europe the German and Allied armies were locked in combat on the bloody battlefield of Verdun. Those first Coast Guard aviators received their wings in time to serve with the American task force in Europe. One, Lieutenant Charles E Sugden, was appointed Commanding Officer of the Naval Air Station, Ile Tudy, France, and was awarded the Legion of Honour by the French government. Coast Guard aviator Elmer F Stone served on the cruiser USS *Huntington*.

After the Armistice the US Coast Guard was returned to the US Treasury Department and the former US Naval air station at Morehead City, North Carolina, was made available as a temporary USCG air station. The US Navy loaned the Coast Guard six surplus US Navy HS-2 flying boats. These aircraft were probably Curtiss HS-2Ls but may have been manufactured by one of the sub-contractors such as L.W.F. (Low, Willard & Fowler), Bolling, Gallauder, Stanard of Loughead-Longbud. After a year this air station was discontinued due to lack of government funds and the aircraft were returned to the US Navy. One can only assume that the Curtiss HS-2L was the 'first' Coast Guard aircraft.

The lean years after World War I were brightened by one major event, this being the race to win the 'Atlantic Blue Ribbon'. The war had spurred the development of aviation, and the idea of an aerial crossing of the Atlantic did not seem as impossible as before. On 17 May 1919, Americans were informed that three US Navy flying-boats, the NC-1, NC-3 and NC-4 were in flight across the Atlantic. Of the three, only the NC-4 completed its journey — the first aircraft to fly the Atlantic. The pilot was Lieutenant Elmer F Stone of the US Coast Guard. The service had helped to win the 'Blue Ribbon of the Air' for the United States of America.

Coast Guard aviation laid dormant through 1922, 1923 and 1924. During 1925 Lieutenant Commander C C von Paulsen, on his own initiative, borrowed a Vought seaplane from the US Navy for a period of one year. The aircraft was possibly a Vought UO-1. A US Army surplus tent was purchased for one dollar. The US Department of Fisheries owned a small island in Gloucester Harbor, Massachusetts, and they allowed von Paulsen to set up a temporary USCG base. This was the famous 'Ten Pound Island' aviation base of the Coast Guard. Due to lack of funding, the base was discontinued and the UO-1 returned to the US Navy.

It was the 'Rum War' of the mid-1920s which again brought Coast Guard aviation into the headlines. It was the turbulent prohibition era when rum-runners with their swift, powerful boats challenged the authority of the US Federal government. The surface craft of the Coast Guard were no match for the smuggler's boats. It was soon apparent that aircraft would have to be brought into the effort.

As a result of a Congressional Act dated 3 March 1926, appropriation to purchase five new aircraft was received and arrangements were made to get them from US Army and US Navy contracts. An initial sum of $152,000 was appropriated. Five seaplanes were purchased and in 1927 two air stations were established, one at Gloucester, Massachusetts, the other at Cape May, New Jersey. Coast Guard aircraft No 1, 2 and 3 were Loening OL-5s while No 4 and 5 were Vought UO-4s. Lieutenant Commanders Elmer F Stone and S S Yeandle of the Coast Guard supervised their construction. Between 1927 and 1930 these five USCG aircraft flew over 200,000 miles. These were the first aircraft that the Coast Guard could call its own, all previous aircraft having been borrowed from the US Navy.

Congressional recognition being secured, and important persons in the Coast Guard being convinced that aviation could play an important role, there was yet the long uphill road of training a group of aviators, first in the handling of aircraft in the more normal sense, and second the application of this new equipment and personnel to the task in hand. Many more difficult problems presented themselves. Indoctrination of service personnel was not the least of these, for without complete integration the new air arm would have only limited value.

Those early years of Coast Guard Aviation included the expected functions of life saving and patrols, plus searching out schools of fish for local fishermen, looking for wrecked seaplanes, and machine-gunning cases of liquor ditched by the rum runners. They were also involved in aerial photography. During 1928 the five Coast Guard aircraft flew 56,395 miles covering an area of 945,275 square miles. No less than 5113 vessels were identified or aided at sea. By 1933, the Coast Guard had 13 aircraft and 14 aviators in active service, and by 1936 the number of aircraft had increased to 45, with 27 aviators on active assignment. US Coast Guard Aviation was definitely established as growth continued and its furture was assured.

Between 1932 and the outbreak of World War II nine new air stations, authorised by the historic legislation of 1916, were built and put into operation. These included Biloxi, Mississippi; Brooklyn, New York; Elizabeth City, North Carolina; Miami, Florida; Port Angeles, Washington State; St Petersburg, Florida; Salem, Massachusetts; and San Diego and San Francisco in California. They were all equipped with seaplane ramps, landing lights, mooring buoys and other facilities required by seaplanes, while the ramp would lead directly to the hangar complex. More than half the air stations had runways so that amphibian aircraft could use land as well as water. Every air station had its radio and communications centre plus aircraft overhaul facilities. There were new barracks and ample recreation areas for the US Coast Guard personnel.

RADIO EXPERIMENTS

During 1925/26 at Coast Guard Base 7 located at Gloucester, Massachusetts, instructional flights were offered and many hours spent in the air for experimental flights in the development of radio communications between aircraft in flight, and between aircraft, ship and ground stations. The original inception of the idea was envisaged by Radio Electrician A G Descoteaux USCG. He constructed the particular type of radio equipment which was installed in the Gloucester seaplane during 1925. It consisted of batteries, head telephones, telegraph key, microphone and accessories, with a two-way continuous wave telegraph and a high-quality voice communication. Designed for battery operation, entirely independent of the aircraft's regular electrical system, and weighing only 90lb, it soon became the basis of the later USCG aircraft equipment.

Cape May USCG Air Station, New Jersey, during 1935, with four aircraft parked on the apron and ramp. They include Douglas Dolphins 'V129' *Adhara*; 'V138' *Bellatrix*; General Aviation PJ-1 'V162' Flying Life Boat, and Grumman JF-2 Duck 'V162'. *(USCG)*

The peculiar nature of the duties of Coast Guard aircraft were such that the regular US Navy and commercial wireless equipments then available were impracticable. Like the patrol cutters, Coast Guard aircraft required a two-way telegraphic and telephonic communication. Efficiency of patrol and rescue operations relied on constant contact with other aircraft, vessels and land stations. Furthermore, long distance communication necessitated the use of highly sensitive receiving equipment, which had to be independant of the aircraft's mechanism, and would ensure ready reporting while on water or in the event of a forced landing. All this involved careful shielding and banding of all the equipment and associated circuits throughout the aircraft. The work of Descoteaux and C T Solt of the Communications Section of USCG HQ resulted in the use of the first loop type radio direction finder. This equipment was adapted to the service and used successfully in Coast Guard aircraft.

We cannot identify the Admiral or his lady, but without doubt the amphibian in the background is General Aviation PJ-1 *Altair* 'FLB-52' which between 1933 and 1935 was based at Cape May, New Jersey, US Coast Guard air station. *(USCG)*

The efficiency of this new radio equipment was demonstrated on 13 June 1929 at Old Orchard Beach, Maine. In a Loening OL-5 amphibian covering the scene of operations, Descoteaux broadcasted the take-off and departure of the 'Yellow Bird' on its historic transatlantic flight. The description of the event by Descoteaux was relayed by ground equipment to an extensive national hook-up. It was clearly received by American radio stations and in several foreign countries. This was apparently the first time in the history of radio that a broadcast of this kind was successfully conducted.

USCG AIR TRAFFIC CONTROL

During 1929 the initial steps toward the formation of a radio communications chain of station hook-ups was introduced for the sole purpose of checking system was inaugurated along the

The launch of a Coast Guard amphibian or flying-boat always stirred local interest, as shown in this photo of PJ-1 *Altair* 'FLB-52' being towed by tractor to the ocean at Cape May, New Jersey. It served here until, as '252' it moved to Salem, Massachusetts in 1936. By 1937 it had become 'V112' and in 1938 was moved south to Miami. *(USCG)*

Atlantic seaboard. Its primary aim was to keep track, by regular radio reports, of all aircraft using the coastal routes. By means of scheduled checks on all departures and arrivals, or check-time when abeam a station, aircraft on long distance flights could be kept under continual radio observation. At all points along the coast from New York to Miami, flight were kept under constant surveillance by radio checks and position reporting.

With no additional cost to the US government, adequate protection and control of airborne traffic was thus guaranteed. When an accident did occur, the nearest Coast Guard station was ready with immediate assistance. During the first two months of operation, 329 aircraft were checked and surveyed. In 1932, statistics showed that some 14,000 reports of transit aircraft were made by the USCG stations scattered up and down the Atlantic seaboard. As more air stations were established, the system was gradually extended to cover the Gulf coastline and the Pacific seaboard. This unique organisation soon became a regular part of the efficient air communications system developed by the US Coast Guard.

It is not generally known that the early Coast Guard aircraft were allocated International W/T (Wireless telegraphy) call-signs as a means of identification. The Vought UO-4s CG4 and CG5 based at Gloucester had the call-signs 'NUMRA' and 'NUMRB', the Consolidated N4Y-1 'NUMRF' and the Douglas RD CG27 *Procyon* 'NUMRG', the latter four aircraft being based at

Three US Coast Guard aviators pose in front of Douglas RD-3 Dolphin 'V129' at Port Angeles USCG Air Station in the late 1930s. *Left to right:* Donald B MacDiarmid, Bill Snyder and Freddy Wild. The RD-2 went to Port Angeles during 1937, being transferred to Salem during 1940. *(USCG)*

Cape May. Miami had on charge Douglas RD-1 CG28 'NUMRH' and two General Aviation PJ-1 Flying Life Boats CG54 and CG55 'NUMRO' and 'NUMRP'. The latter two were named *Acamar* and *Arcturus* respectively. Gloucester also had Douglas RD-2 CG29 *Adhara* 'NUMRJ' while Cape May operated a trio of General Aviation PJ-1s CG51, CG54 and CG53, 'NUMRL', 'NURM', 'NUMRN' named *Antares, Altair* and *Acrux*.

Today radio communications and frequencies have advanced so much with the efficient avionics packs and black boxes, enabling both US Coast Guard vessels and aircraft to monitor any given radio frequency at any given time.

EARLY USCG ACTIVITIES

When the Morehead City, North Carolina, air station was opened on 24 March 1920, the Coast Guard assumed custody of the US Naval air station buildings there. The six aircraft procured from the Bureau of Aeronautics on temporary loan included, in addition to the Curtiss HS-2L flying boats, two biplane types known as Aeromarine 40 powered by Curtiss OXX engines. They were retained until the station was decommissioned, but being surplus wartime aircraft they were then destroyed as unsuitable for further duty. Although no funds had ever been appropriated for its maintenance, the air station prospered during the first fifteen months of its history. Finally, on 1 July 1921, USCG HQ was forced to put it on the inactive status due to lack of financial support.

General muster at Charleston, South Carolina, USCG Air Station during 1940. The aircraft positioned behind include Douglas RD-4 'V132' and Grumman JF-2 Duck's 'V140', 'V141' and 'V146'. The station also had on strength at this time two more RD-4 Dolphins 'V125' and 'V126'. *(USCG)*

This had been the first real attempt at the use of air patrols for Coast Guard duties so the Commandant — W E Reynolds — set forth the particular functions of the aviation programme. Seven distinct tasks were enumerated, which, in addition to the regular Coast Guard duties for all units, comprised experimental flights, including reconnaissance of land water areas, surveying, mapping, routes for lines of communications, assistance in flood control and relief work for the western rivers region, plus emergency transportation services for government officials to remote or inaccessable locations.

Chiefly, the air station activities were in the enforcement of Federal maritime laws, patrol duty, and general humanitarian

A single Martin T4M-1 torpedo scout bomber appeared in the USCG aircraft inventory. It was delivered to the Engine School & Repair Base at Norfolk, Virginia, on 2 June 1937. It had the US Navy BuNo A-7607 and was used for ground instruction. Depicted is a US Navy T4M-1 BuNo A-7596.
(Peter M Bowers)

work along the coast. The Coast Guard annual report for 1920 indicates however, that during the course of the year's work, the aircraft performed valuable services of a miscellaneous nature. Many vessels and wrecked aircraft were assisted, fishermen were aided in various ways, several reconnaissance, surveying and mapping flights were made, and numerous individuals afforded transportation. On two different occasions, aircraft were despatched a distance of over 60 miles to carry a physician to isolated spots to attend to persons in need of special attention. It was a beginning. Regardless of the restricted operations of the station, it had proved that aircraft had a definite role in all future US Coast Guard expansion.

When Lieutenant Commander von Paulsen secured the loan of a Vought UO-1 from the US Navy in 1925 at Gloucester, Lieutenant L M Melka flew with him as pilot and mechanic for the aircraft. It was fortunate indeed that this pioneer air station had the services of two of the Coast Guard's ablest aviators to direct its early ventures. Until a move to Ten Pound Island procured from the US Bureau of Fisheries, for several months the UO-1 was housed with the US Naval Reserve air station at Squantum. Von Paulsen and Melka alternated in regular patrols, making one to three patrol flights every day during 1925/26. The smuggling of all sorts of contrabands was rife.

On 8 July 1940 two Great Lakes BG-1 bombers were delivered to the USCG Engine School & Repair Base, Norfolk, Virginia. They were ex US Navy BuNos 9506 and 9519, with 1433 and 1611 flying hours respectively. They were used for ground instruction. Depicted is a BG-1 BuNo 9840 of the US Marine Corps seen on 14 October 1935. *(Peter M Bowers)*

A Grumman J2F-5 Duck is seen stowed securily on board the USCG cutter *Storis* off the east coast of Greenland on 25 August 1945. Constructed during World War II, the cutter was never intended to carry aircraft. The Duck was winched in and out of the water, before and after flight. *(USCG)*

Meanwhile, other air stations were being commissioned under the provisions of the 1916 Act. In 1926, the second main air station had been activated at Cape May, New Jersey, as mentioned earlier. Increased smuggling of aliens and liquor along the Florida coast in 1928 led to the temporary assignment of two USCG aircraft at Fort Lauderdale, Florida, as an adjunct to the Coast Guard section base there. Up until 1932, the Gloucester and Cape May air stations were the only permanent air bases for the fleet of aircraft which was growing steadily. In 1933 however, a new air station was commissioned at Dinner Key, Florida. The Coast Guard annual report indicates that the site was leased in 1932 and, after a hangar had been completed, the station was commissioned during the following year. This equipment apparently became part of the Miami air station, since no further mention of the Dinner Key air station is made in any later reports.

On 9 March 1934 the US Treasury authorised the consolidation of all its aviation activities. Accordingly, fifteen miscellaneous aircraft of the US Customs Service were turned over to the Coast Guard. All were in poor condition, and only two New Standard aircraft subsequently went into service with the USCG. By January 1935, these two New Standard NT-2 aircraft 311 and 312 were based at the air station at Miami. At this time three new air patrol detachments were established at Buffalo, New York; and San Antonio, Texas, where a Vought O2U-2 301 was based, and San Diego, California, with a Grumman JF-2 167. Six landplanes were transferred from the US Navy to these bases to aid combating smuggling activities across both the Canadian and Mexican borders.

Photos such as this depicted are always of interest to USCG personnel — recognise anyone? General muster at Elizabeth City on 25 May 1950, depicting many veterans of all ranks. Seen parked behind the assembled ranks are two Martin P5M- Mariners, plus a partially hidden Sikorsky HO3S-1G helicopter. *(USCG)*

To absorb new entrants to Coast Guard Aviation, an aviation training school for elisted personnel was founded at Cape May, New Jersey. During 1935, when the Salem, Massachusetts, air station was commissioned, the Gloucester base was put on the inactive list. Cape May air station, however, continued until 1938. During 1935 further special air patrol detachments were established at Charleston, South Carolina, and, from time to time, at strategic points along the coast. The Charleston detachment was commissioned as a permanent air station in 1937, being equipped with a unique selection of USCG aircraft which included a General Aviation PJ-1 V114, a Douglas RD-4 amphibian V125, a Grumman JF-2 V147, a Viking OO-1 seaplane V156, and a Fairchild J2K-1 landplane V161. This air station continued in active operation until 1942.

Scene at the USCG Rescue Control Centre (RCC) in New York on 3 March 1953, as Captain Theodore J Harris and Lieutenant George A Gyland of the Coast Guard plot the position of a ditched Lockheed P2V- Neptune of the US Navy, in the Atlantic four days earlier. *(USCG)*

Four of the nine Consolidated PB4Y-2 Privateer patrol aircraft acquired by the USCG are seen flying over the Golden Gate Bridge, San Francisco, on 11 March 1958. Aircraft '6260', '6300' and '6260' were based at San Francisco CGAS during the 1950s. *(USCG)*

Sunday 31 January 1960, Sikorsky HUS-1G '1332, '1334' of the USCG Air Detachment, New Orleans, Louisiana, and its USCG personnel commanded by Lt Cmdr James H Durfee and Lt Cmdr James L Sigma, Executive Officer. We have the names of the rest of the personnel if anyone is interested. *(Sikorsky)*

The original five aircraft of 1926 were increased to one short of fifty by 1941. There were ninety pilots in 1941, who were all graduates of the Naval Air Station at Pensacola. However the USCG archives indicate that other enlisted men were trained at the US Army aviation school located at Mineola, New York, prior to World War I. The total of the small Coast Guard aviation group was eighteen at the time of the US entry into the conflict in 1917. On 1 November 1941, the Coast Guard was moved from under the jurisdiction of the US Treasury to the US Navy Department. This transfer and the outbreak of World War II placed entirely new demands upon US Coast Guard Aviation.

The US Coast Guard inventory of 1 June 1965 indicates that Salem air station, Massachusetts, was equipped with four HU-16E Albatross amphibians — '1242', '1266, '1293' and '7226'. Depicted is the air station personnel posed in front of HU-16E '1242'. *(USCG)*

Coast Guard Aviation activities during 1940 with fifty-three aircraft on the inventory were distributed between ten air stations, plus a single Grumman JF-2 V135 on the cutter *Taney*, based in Honolulu, Hawaii. Up to 1939 a Curtiss SOC-4 V172 had been based on the Bibb at Norfolk, Virginia. Aircraft located 725 illicit distilleries, warned 1466 persons of impending danger, assisted 223 persons, transported 113 emergency medical

This USCG Air Station at Astoria, Warrenton, Oregon, was established on 25 February 1966 after being at Tongue Point Naval Air Station for over two years. In March 1973 HH-52A helicopters were replaced by three HH-3F Pelicans. This aerial view of the air station was taken on 7 August 1974 with a HH-3F parked on the apron. *(USCG)*

cases, and airlifted twelve persons from disabled vessels, in addition to identifying no less than 29,322 vessels. The Coast Guard aviators and aircraft completed 4801 fights, cruising 1,258,344 miles, covering an area of 9,307,066 square miles resulting in 13,231 flight hours.

Today, seventy-five years after the formation proper of US Coast Guard Aviation in 1926, one must recall that Coast Guard involvement in aviation extends back as early as 1900 when personnel from the US Life Saving Stations in the vicinity of Kitty Hawk, North Carolina, became involved in the experiments and activities of a certain pair of bothers whose normal occupation was the manufacture of bicycles in Dayton, Ohio.

The dedication of any new USCG facility is always a chance to publicise the service. Hangar scene on 28 September 1974 at North Bend, Oregon, with US Congressman Dellenbach at the podium, at the dedication of the new Air Station. In the background is a HH-52A '1369'. Including the USCG Group facilities, the air station cost $2.4 million. *(USCG)*

Those surfmen provided logistic support for the Wright brothers, and assisted routinely in the launching of their experimental gliders. These men, members of one of the agencies which were later combined to form the modern day US Coast Guard, were already classed as a skilled ground crew when they were called upon to assist in the memorable events of 17 December 1903.

Since 1965 US Coast Guard Aviation has presented the Ancient Albatross award to the Coast Guard aviator on active duty holding the earliest designation in recognition of a clear defiance of the private realm of the Albatross and all its seabird kin while in the pursuit of time honored Coast Guard duties.

Elmer F Stone would surely have been proud of US Coast Guard Aviation today, as would many other USCG pioneers of the early meagre days such as Hall, Chiswell, Melka, Eaton, Coffin, Parker, Donohue, Wishaar, von Paulsen, Descoteaux, Solt and Yeandle. We salute you . . .

A USCG crew-person hastens from a Sikorsky HH-52A Seaguard from Cape Cod, Massachusetts, Air Station, across a stony beach to the aid of a person in distress during July 1979. Females fly both helicopters and fixed-wing aircraft, including the HC-130H Hercules, with the Coast Guard. *(USCG)*

Seen parked, during landing trials, on the new USCG cutter *Alert* is Aérospatiale HH-65A '6582' acquired by the service on 27 September 1988 and delivered to the Air Training Centre, Mobile, Alabama, from where ship-borne trials were conducted. *(USCG)*

CHAPTER TWO
US Coast Guard Aircraft Markings and Insignia

The colouring of US Coast Guard aircraft has changed considerably over the years. In the history of US Coast Guard Aviation commencing with World War I, aircraft appeared in varying paint schemes and really did not become uniform until after World War II. Until 1936, US Navy colours were used, with the vertical red, white and blue stripes on the rudder. Contemporary US Navy dark blue hull and yellow wing may be seen in

This view of the only Douglas O-38C observation land plane 'CG-9' to serve with the USCG shows clearly the blue fuselage and yellow wing and markings. It initially served at Cape May, New Jersey, and used the International radio call-sign 'NUMRD'. It crashed during April 1934.
(Peter M Bowers)

illustrations of Coast Guard OL-5, UO-4, RD, RD-1, RD-2, O-38C, and the PJ amphibians. During 1936, to provide a more distinctive colour scheme, the aircraft paint scheme was changed to aluminium, with yellow upper wing and tail surfaces with a red hull and white rudder stripes topped by a blue field. Some USCG aircraft in service were repainted. This colour scheme is evident in illustrations of the OO-1, NT-2, RD-4, R3Q-1, RT-1 and J2W-1.

The years 1927 to 1936 have been described as the Coast Guard's 'blue and silver' period. From the first Loening OL-5 until the instruction of the 'V' serial system in 1936, every aircraft had a distinct colour scheme. The underside of the hull, on flying boats, was painted silver, with the FLB-type wing floats being all silver, and the Douglas Dolphin's half silver and half blue. On flying boats the engine nacelles were silver and the

This Consolidated N4Y-1 was commissioned by the USCG during August 1932, initially being based at Cape May, New Jersey, where its serial was changed from 'CG-10' to '310' and then 'V110'. During 1938 it was moved to Brooklyn, New York, and the following year to Biloxi, Mississippi. By 1941 it was based at Elizabeth City CGAS, North Carolina. *(Peter M Bowers)*

cowls blue. The name US COAST GUARD was painted in large letters on the side of the fuselage with the letters USCG on the underside of each wing. The letters US were on the top of the port wing, and CG on the top of the starboard wing.

From 1936 to 1941, to provide a more distinctive colour scheme, all aircraft were an overall aluminium colour with the top of the wing chrome yellow along with the upper part of the tail surfaces. Commencing in late 1935 the top one-third of the rudder was painted insignia blue, the bottom two-thirds being divided into five equal vertical stripes, three red and two white. The aircraft's model designation was placed in white on the blue section of the rudder, and the serial number was in black under the words US COAST GUARD on the side of the fuselage. The letters USCG were used on the underside of both wings, and the serial number was painted on the bottom of the hull or fuselage. The USCG emblem was painted on the forward part of the hull or fuselage near the pilot's compartment.

This early photo is of Douglas RD-2 Dolphin with serial 'CG-29' and named *Adhara* and showing clearly the USCG letter markings on top of the yellow wings. During 1933 this Dolphin was based at Gloucester Air Station in Massachusetts, and used the International radio call-sign 'NUMRJ'. *(Peter M Bowers)*

The early days primarily rested upon amphibian models, since the necessity of water landings was always uppermost in the mind. The first five USCG aircraft included a trio of Loening OL-5 amphibians, essentially modification of a US Navy design, but tailored to a derivative with a stronger fuselage and increased fuel capacity. During this period of aviation's romance with flying boat/amphibian craft it was always a moot point whether it was a boat with wings added as an afterthought. The ultimate aim from the beginning of Coast Guard aviation had been a 'flying life boat' and the theory and even deep research continued until well after World War II.

When US Coast Guard and its aviation was transferred to the US Navy commencing its wartime operations under the Secretary of the Navy, in accordance with the law and an

Excellent photo of a General Aviation PJ-1 'FLB53', *Acrux* '253' showing clearly the markings of the period 1935/36 when it was based at Cape May, New Jersey. Blue fuselage, silver under-surface, orange/yellow tops of wing and tail. Tail stripes were red/white/blue — red forward. The USCG emblem is under the cockpit. *(Gordon S Williams)*

By 1937 the PJ-1 FLB-53 had been correctly re-designated, repainted, and re-registered 'V113'. It is seen beached by the ocean at Cape May, New Jersey. By 1938 it had been assigned to Brooklyn, New York. No US national insignia appeared on USCG aircraft until taken over by the US Navy after the attack on Pearl Harbor. *(Gordon S Williams)*

executive directive dated 1 November 1941, the yellow tail band appeared on aircraft assigned to air/sea rescue duties. Later, US Navy colours were applied with grey and blue grey being commonly used and some aircraft appearing in a species of camouflage. Red and white horizontal tail stripes applied to US Navy aircraft between January and May 1942 were also used on some US Coast Guard aircraft.

After World War II the US Coast Guard was returned to the US Treasury Department taking with it a mixture of aircraft still with USCG 'V' serials and US Navy aircraft with BuNos. The letter 'V' was dropped from the few original Coast Guard aircraft and on the first day of January, 1951, the numeral '1' was added as a prefix to make up a four-digit number for use as a radio call-sign. Until 1947 the standard location of the serial number remained near the top of the vertical fin. It was then moved to a point on the fuselage below the leading edge of the horizontal tail. The serial number, later to become a varied mixture of US Navy and US Air Force was usually in figures a little larger than those used for the model designation and name of the service. Later, and until quite recently, the last four digits of the USCG serial appeared on the nose of the aircraft in figures twelve or more inches high.

The wartime two-tone camouflage disappeared, all aircraft reverting to a general pre-war scheme, being either natural metal finish or painted aluminium. Helicopters were given an overall yellow/orange finish. The name US COAST GUARD was painted on the side of both fixed wing and rotary wing aircraft. The national star and bar insignia was placed on the top of the left wing and the bottom of the right, and on both sides of the fuselage. The letters USCG were on the top right and lower left side of the wings. On helicopters caricoloured tips appeared on

Except for the World War II years, the Coast Guard intended that identity of their aircraft would be simple, adequately marking the underneath of all types in the pre-war years. Depicted is Douglas RD-4 'V128' from the San Francisco CGAS seen over the Oakland estuary in California during May 1941. *(Peter M Bowers)*

rotating aerodynamic surfaces. This was to enhance visibility to USCG personnel working around helicopters on the ground, and helped track the main rotor blades.

All aircraft types considered to have the primary task or mission of Search & Rescue (SAR) had additional high visibility markings. The upper and lower portions of the wing tips — approximately seven per cent of the wing span — were painted chrome yellow with an additional six inch black band inboard. An orange/yellow band encircled the fuselage, eighteen inches

Factory-new Douglas RD-4 Dolphin '130' photograph on 11 March 1934 prior to delivery. On entering USCG service it was named *Spica* and served at Cape May CGAS, New Jersey, until 1936. By the time it had moved to Charleston, South Carolina, in 1937 it had been re-registered 'V125'. *(Douglas SM-7010)*

wide of the Grumman JRF Goose, and thirty-six inches wide on all other aircraft. This band was enclosed by a three or six inch black band respectively. The new colour scheme was introduced during December 1952. This new scheme could be seen on the PBY, PBM-5, P5M-1G, P5M-2G and the HO3S-1G helicopter. As mentioned earlier the yellow tail band was actually applied during World War II to both US Navy and US Coast Guard aircraft assigned to air/sea rescue, some aircraft like the PBY, and P5M, carrying the word RESCUE in black on a yellow panel background on the top mainplane.

According to the USCG aircraft register, this Curtiss SOC-4 Seagull landplane 'V171' was based at San Diego CGAS, California, during 1938, moving to Miami in 1940 where it remained during 1941. It also did a tour equipped with its floats on board the USCG cutter *Bibb*. *(Gordon S Williams)*

HIGH VISIBILITY

During 1958, Coast Guard aeronautical engineers began experimenting with a new easy visibility paint scheme. Norman N Rubin, US government aeronautical engineer employed in Coast Guard headquarters, was involved in these studies which led to the adoption of the red splash. Two disparate problems converged on his desk that year. The first was the 'see and be seen' campaign being waged by the US Department of Defense in the aftermath of a series of mid-air collisions and near misses with airliners. Emphasis upon increased flying discipline was only part of the answer. Because of the widespread use of aluminium as the bare skin of aircraft, the most alert crews could be caught by surprise. Silver colour makes a good camouflage, except in the unusual case of sunlight glinting

Expertly restored to flying condition by Bob van Vranken of Saisun, California, is this USCG N3N-3 '2582' often demonstrated at air shows. The aircraft was discovered by Bob in a cropduster's boneyard at Fresno in 1970, but never converted. Bob discovered he had actually trained on this N3N-3 at Glenview, Illinois, in 1940. *(Jay Wright)*

Unusual photo of a large Hall PH-3 flying boat taken in the early years of World War II showing the US national insignia replacing the USCG fuselage serial. The service was then under the jurisdiction of the US Navy. Records reveal that this PH-3 could be 'V177' which in 1941 was based at Brooklyn, New York. *(USCG)*

off it. Several complex lighting systems were proposed to provide instant information, none of which achieved widespread acceptance. Cornell University assisted in research and suggestions were obtained from other interested agencies. The idea was to paint aircraft conspicuously as a preventative to mid-air collisions, which was becoming the concern of the whole aviation industry, the military services, and a jittery and nervous public.

There is a good chance that this USCG Douglas R4D-5 BuNo 17183, depicted in 1961 whilst in storage at the Davis-Monthan depot, could still be a survivor. It became US civil registered N2204S in 1964, and in 1984 was sold to Bolivia and registered CP-1940 so could be still flying. *(MAP)*

The most promising scheme was a fluorescent blaze orange outlined in black appearing on the nose of the aircraft, in a band around the fuselage behind the wings, and on the tail, on a basic field of solar heat reflecting white paint. These experimental paints proved many times more costly than the regularly used lacquers and enamels. They had to be stripped and replaced three or four times a year for maximum efficiency. Ultraviolet rays of the sun damaged the organic dyes so that the paints faded and lost power. Grumman HU-16 amphibians were the basis of these interesting experiments, and applied to Albatross aircraft based at the San Francisco station in California.

After extensive further research easily strippable and more durable paints were developed and used on Coast Guard aircraft by 1960. A new paint scheme was incorporated in 1967, basically similar to the earlier design, but incorporating the new diagonal service identification blue bordered red stripe with the US Coast Guard emblem imposed on it.

Fluorescent paints that were available offered promise. It seemed wise to limit use of these paints to small areas of the aircraft and so avoid a dazzle effect. In 1958 several schemes were considered, and finally Norman Rubin evolved an arrowhead design over the nose of the aircraft in an attempt to give an indication of the direction of motion. It was necessary, of course, to keep the markings within the framework of national and international usages for state owned law enforcement and Search & Rescue aircraft. The geometry of the Grumman Albatross aircraft, a Coast Guard mainstay of the period, made the application of an arrowhead awkward, and it became a slash, leaning forward at the top. Sometime later, a small blue stripe was added for cosmetic effect.

When the 'V' serial system was adopted in 1934, it was retained on some aircraft well into World War II, being cancelled on 28 December 1945. This resulted in some aircraft procured by the USCG during the war had a 'V' added to the US BuNo or US Air Force serial. A good example is this Boeing PB-1G Flying Fortress with the tail serial 'V 77256'.
(Gordon S Williams via Museum of Flight)

The second problem was offered by the US Coast Guard's Public Information personnel. Their need was to so mark the Coast Guard aircraft that their identity could not be missed in press publicity photos. Of course the legend US COAST GUARD appeared in letters two feet high on both sides of fuselages and hulls, but the USCG public information officials showed Norman Rubin a photograph depicting a helicopter rescue operation in which the Coast Guard legend had been painstakingly retouched to identify the helicopter as belonging to another service.

A statistical study of photographs of people deplaning after search and rescue missions indicated that the highest possibility of a legend appearing in such photos was offered by locating the words either at left or above the cabin exit. The smallest letters legible in typical newspaper halftones would be three

In the early post-war years USCG markings on types acquired from other US services were far from standardised. This Douglas R5D-4 BuNo 5490 had Arctic red tail and nose, a red fuselage band and red outer wing panels. Note the position of the US national star and bar insignia. It is seen at San Francisco on 27 September 1959. *(William T Larkins)*

inches high. This size and location were adopted. For long-range photos, the fluorescent orange bow bordered by a black slash, coupled with a bold legend and a Coast Guard insignia of good size, made the aircraft uniquely recognisable as Coast Guard. As a final touch, the overall colour of white was adopted, both for uniqueness and to satisfy service demands for solar heat reflection to keep down cabin temperatures in the summer sun.

Unusual photo taken on 30 January 1951 showing clearly the markings used on the Coast Guard's one and only Kaman K-225 helicopter. All details are contained on the rear stabiliser. It is depicted parked on the apron at Elizabeth City CGAS, North Carolina. *(USCG)*

There were exceptions to the rule and additional means of identification. Helicopters which were seconded to US Coast Guard cutters deployed to the Arctic and Antarctica regions were given a deep red overall finish for ease of identification in snow and ice areas. Up to the drug interdiction conflict the US Coast Guard air station ident or name appeared on all aircraft, and during the last days of the ubiquitous Grumman HU-16E Albatross, the air station ident was painted on the inner side of the wing floats so that press publicity photos taken from the fuselage hatch automatically included the wing float and air station name. Sikorsky HH-3F Pelican helicopters based at San Francisco air station in California carried a Golden Gate Bridge image above the pilot's cabin in front of the rotor head.

Nose art is seen on rare occasions on US Coast Guard aircraft but a Sikorsky HH-52A Seaguard helicopter from the USCG Polar Operations Division (AVTRACEN) based at Mobile, Alabama, carried a cartoon character on its nose similar to 'Yosemite Sam' and the name 'Whirleybird Cowboy'. Another HH-52A seen at Mobile had just returned from detachment on an icebreaker and featured nose art of the 'Tasmanian Devil' cartoon character. During the period of the return of Haley's Comet a HH-52A from the Polar Operations Division was deployed in Antarctica so the crew was in one of the best locations to observe the comet. Artwork on the helicopter included a comet superimposed on the continent and surrounded by the Southern Cross. Coast Guard helicopters on deployment

Accepted by the US Coast Guard on 19 March 1963 this Grumman VC-4A Gulfstream I '1380' is seen flying over snow covered Virginia terrain on 16 January 1964. It carries ARLINGTON on the tail, the name of the USCG Air Detachment. It was later repainted and given the serial '02'. *(USCG)*

Flagship of the US Coast Guard Aviation fleet is this immaculate Grumman VC-11A Gulfstream II '01' high-speed executive jet transport used by the USCG Commandant. It is based at Washington National airport. The Coast Guard Air Detachment Arlington was commissioned on 20 February 1952. *(Peter M Bowers)*

to cutters and icebreakers often carried the ship's name in addition to the home air station name. An example was a HH-52A which carried *Polar Sea* on the tailboom and Mobile on the cargo door below the national insignia. The *Polar Sea* is one of the Coast Guard's icebreakers. Colour markings are very important to the model maker, and a fine selection of USCG aircraft colour photos are to be found in *A History of US Coast Guard Aviation* published by Airlife.

Seen at Opa Locka airport, Florida, is Convair HC-131A '5783' from the local Miami CGAS. Fourteen of the eighteen aircraft procured came from the Davis-Monthan storage depot, the remaining four from the Air National Guard. Rework and modification was conducted at the Aircraft Repair & Supply Centre, Elizabeth City, North Carolina. *(USCG)*

Helicopters of the USCG seconded to cutters deployed to the Arctic and Antarctic regions were given a deep red overall finish for ease of identification in snow and ice areas. Depicted is Sikorsky HH-52A '1428' from Detroit, Michigan, Air Station sporting the red finish. *(AP Photo Library)*

On 18 June 1951, an excellent painting and instructions Memorandum No 7-51 was issued by the USCG Engineer-in-Chief. This twenty-four page document was most detailed in every respect and included the following aircraft: Bell HTL-1; Boeing PB-1G; Beechcraft JRB-4; Douglas R4D-5; Douglas R5D-3/4; Consolidated PBY-5AG; Grumman JRF-5G; Grumman UF-1G; Lockheed R50-4/5; Martin PMB-5G; Piasecki HRP-1G; Sikorsky HO3S-1G; Sikorsky HO4S-1G; Sikorsky HOS-1G and Stinson OY-1.

CHAPTER THREE
The US Coast Guard Aircraft Serial System

The first ever aircraft designed for air/sea rescue work was a triplane flying boat, its control surfaces mounted high on the tail boom, designed and built by Glenn Curtiss. The Curtiss F Boat trainer, in which the pioneer US Coast Guard aviators, Second

The first General Aviation PJ-1 'FLB-51' *Antares* was modified in 1933 and re-designated PJ-2. The engine nacelles and cowlings were reversed, and the cockpit and other sections modified. It is depicted as 'V116' and served between 1937/39 at Biloxi USCG Air Station, Mississippi. *(USCG)*

Lieutenant B Hall and Third Lieutenant Elmer F Stone, from the USCG cutter *Onondaga*, first flew in at the Curtiss Aeroplane and Motor Company School at Newport News, Virginia, was one of the first successful flying boats developed, and it provided the practicability of over-water search by aircraft.

Other flying boats were borrowed by the Coast Guard from the US Navy. By 1940 the service had no less than fifty aircraft operating from twenty-eight air stations. In that time no less than three separate numbering or serial systems had been used by the Coast Guard for their aircraft, mainly procured direct from the manufacturers to fit a USCG specification, the exception being those aircraft later procured from the US Navy and the US Air Force, which were used during and shortly after World War II.

Prior to and after the war US Coast Guard aircraft were always very distinctive in their colour schemes, and it is only over the past twenty years that the markings have become more distinct and standardised. The older US Coast Guard types are now graciously retired, these including the Grumman

Douglas RD-4 '134' *Deneb* just airborne from Port Angeles CGAS, Washington, where it was based during 1935/36. It carried the Douglas Aircraft Company motif on the fin and was c/n 1271. Ten Dolphins were purchased by the service on a US Navy contract for $60,000 each. *(Gordon S Williams)*

HU-16E Albatross amphibian and the Sikorsky HH-52A Seaguard helicopter. Types on the inventory and in service include the HH-65A Dolphin, the HU-25 Guardian, HC-130H Hercules, these three being the mainstay of the service, supplemented by other aircraft.

THE EARLY DAYS

Because so little has been published until recently on the history of the US Coast Guard, and especially its aviation, there can be little wonder that the correct overall picture of USCG aircraft has never been fully understood, especially when one realises that the aircraft serial numbering system, or identification numbers, have undergone no less than five complete changes

Always of interest to modellers are close-up photos. This unusual shot depicts a Hall PH-2 flying boat 'V169' from Port Angeles CGAS, Washington, where it was based during 1938/39. In 1940 it moved to warmer climes to Miami CGAS, Florida, and was still there in 1941. *(Gordon S Williams)*

over the years. This has resulted in some aircraft carrying as many as three different serials — at different times of course — others two, and some only one. For example, the Douglas RD-1 named *Sirius* was at first numbered '28' later becoming '128' and still later V109. The General Aviation PJ-1 named *Altair* was first numbered '52' then '252' and later V112 and in addition was known and marked as 'FLB-52' (Flying Life Boat 52). Photos of Coast Guard PJ-1s have a possibility of showing fifteen different serial numbers, although there were only five aircraft ever in use at one time.

During 1927 a system was adopted involving both a one and

This trio of Grumman JRF-2 Goose amphibians did not appear on the Coast Guard register until 1941, with 'V175' and 'V186' being delivered to Elizabeth City CGAS, and 'V185' to Miami, Florida. This Grumman photo depicts the three prior to delivery in 1940, with 'V185' and 'V186' remaining in service until 1946. *(Grumman)*

Grumman J4F-1 Widgeon 'V198' c/n 1223 was commissioned with the USCG in July 1941, remaining in service until November 1946. During the early post-war years it was based at the Coast Guard Air Station at Floyd Bennett Field, Brooklyn, New York. It is depicted flying over the Brooklyn skyline. *(Grumman)*

two-digit serial on only sixteen aircraft which were broken up into selective groups so that the Douglas Dolphins fell into the '20' serial system and the General Aviation FLBs into the '50' series.

In 1934 the Coast Guard adopted a new three-digit grouping series: '100' for amphibians, '200' for flying boats, '300' for landplanes and '400' for convertible designs that could operate from air stations. Two years later on 13 October 1936, all US Coast Guard serial numbers were completely changed. This was a sweeping revision which included not only all aircraft in service at the time, but for administrative purposes every aircraft that the US Coast Guard had ever flown, regardless of whether it may have already been destroyed or disposed of. Thus, the first ever Coast Guard registered aircraft, a Loening OL-5 CG1, became V101 under the new system. The 'V' system was consecutive and complete with no numbers being omitted, as had been the case with the previous two systems, and this stayed in effect for ten years.

After the US Coast Guard combined with the US Navy on the first day of November 1940, their aircraft retained the 'V' serial but US Navy aircraft attached to the Coast Guard, and flown by USCG aviators, retained their US Navy BuNos. They were engaged in convoy protection and escort duties, anti-submarine warfare (ASW) duties, plus routine patrols and rescue duties. When all Coast Guard equipment, including aircraft, returned to the US Treasury Department on 30 June 1946, it had a varied mixture of aircraft with 'V' serials and US Navy BuNos. Because the US Navy serials were sometimes incorrectly referred to as Coast Guard numbers, it was decided that with effect from 28

Lockheed R50-4 BuNo 12453 seen parked at Elizabeth City USCGAS, North Carolina. Commissioned during 1942, it was based at Brooklyn, New York, in 1943. The Lodestar was used for administrative flights before the Coast Guard Air Station was established at Washington National Airport near Coast Guard headquarters. *(USCG)*

Unusual view of a Coast Guard Sikorsky HUS-1G '1382' hoisting a patient in its rescue basket on 14 December 1959. It was based at the USCG Air Detachment located at New Orleans, Louisiana, and the basket and hoist saved many lives in the myriad bayou areas around the Mississippi River Delta. *(USCG)*

December 1945, the prefix letter 'V' was to be deleted from all Coast Guard aircraft. This ended a very unique and fascinating period of US Coast Guard Aviation history.

POSTWAR YEARS

There was the odd occasion when a US Coast Guard aircraft obtained during World War II from the US Navy for some obscure reason had a 'V' prefix added to the US Navy BuNo. Examples were a PBY Catalina and a Martin PBM Mariner. On 1 January 1951, the numerical '1' as added as a prefix to existing Coast Guard aircraft, replacing the earlier 'V' and marking a four-digit number for use as a radio call-sign. At the same time, any five-digit US Navy BuNo on Coast Guard aircraft were reduced to their last four digits, to comply.

The standard location of the serial number remained near the top of the vertical fin until 1947 when it, along with the model and service designation, was moved to a position on the fuselage below the leading edge of the horizontal tail. The serial number was usually in figures a little larger than those used for the aircraft model designation and name of the service. Today, on the HC-130H Hercules, the HU-25 Guardian, and the two VIP Gulfstreams, the four-digit US Coast Guard serial (two in the case of the Gulfstreams) normally appears on the nose of the aircraft in figures twelve or more inches high. The serial on the HH-65A Dolphin helicopter appears on the tail, while on the HH-3F Pelican it is on the tail boom, with the last two numbers

Still playing a major and vital role in Coast Guard operations is the ubiquitous Lockheed HC-130 Hercules. Depicted is the first one '1339' delivered on the last day of 1959, seen flying along the Californian coastline on 7 March 1971. It was finally retired in April 1986. By late 1990 the service had no less than thirty-one Herky Birds on inventory. *(USCG)*

By July 1971, this Fairchild C-123B Provider '4540' was based with the Coast Guard at the Air Detachment base on Guam Island in the Marianas in the Pacific Ocean. The type was based also at the Air Detachment at Naples, Italy, and at the CGAS Miami, Florida, being phased out later during 1971. *(AP Photo Library)*

appearing below the rotor head. On the new Sikorsky CH-3H helicopters the last four numbers of the US Air Force serial appear on the outboard sponson, and last two digits below the rotor head. However, a low profile restriction on air operations, not only in the drug war areas, but overall, has involved the removal, in some cases, of the nose serial numbers plus any air station name and other identification.

Under a Department of Defense directive dated 6 July 1962, all aircraft designations allocated to the US Navy and US Coast Guard were standardised with those of the many US Air Force variants. It was September 1962 before Coast Guard aircraft designations were changed into the new tri-service designation

Rare photo depicting a Cessna 500 Citation 'CG-519' and Israeli Commodore 1123 'CG-160' parked on the apron at Washington National airport on 24 April 1973. Both were involved in a six month evaluation exercise between 1 April and 30 September 1973. *(USCG)*

A Grumman HU-16E Albatross amphibian '7213' flying a law enforcement patrol in the North Atlantic on 2 November 1978, making a pass over a Japanese longline fishing vessel. The *Goat* as the HU-16 was called, was based at Cape Cod, Massachusetts, USCG Air Station. *(USCG)*

system which for a time caused great confusion. Thus, the Grumman UF-2G Albatross amphibian became re-designated HU-16E.

Two oddballs in the unique US Coast Guard serial numbering system, which are still in service, are the two Grumman Gulfstreams based at Washington National air station. The Grumman VC-4A was acquired in 1963, orginally being allocated the USCG serial '1380'. In February 1969 a VIP serial sequence was adopted when they new Grumman VC-11A Gulfstream 2 was acquired, it becoming '01' and the older Grumman VC-4A becoming '02'.

Secondary mission roles of the HH-65A include patrol and observation, passenger transport, and cargo hook operations. Dolphin '6587' was acquired by the USCG on 3 November 1988, its first assignment being to Cape May CGAS, New Jersey, during December 1988. *(Robert F Dorr)*

Seen parked at Oshkosh, Wisconsin, on 28 July 1990 is this immaculate Grumman HU-16E '7226' 'San Francisco', now restored after retirement, and registered N226CG. It is owned and flown by 'Connie' Edwards of Big Spring, Texas, who is a Hon Member of the USCG Pterodactyls. The Albatross attends many USCG Aviation functions within the USA. *(Frank A Hudson)*

Deep research indicates that due to the US Coast Guard being placed under the operational control of the US Navy on the first day of November 1941, the service flew a mixture of aircraft types from both services. Unfortunately these are not all identified. The extensive Grumman history records indicate that the Coast Guard obtained a Grumman J2F-3 unarmed aircraft fitted out in luxury as an 'admiral's badge' but this aircraft was not assigned a USCG serial number, retaining its US Navy BuNo.

US Coast Guard Aviation has travelled far since those two pioneers, Norman Hall and Elmer Stone, took a flight in the borrowed Curtiss F flying-boat over Newport News harbour, a dozen years after the Wright brothers succeeded with heavier-than-air flight at Kitty Hawk.

EARLY US COAST GUARD AVIATION SERIALS

1/2-Digit	Type	3-Digit	'V' serial
1	Loening OL-5		V101
2	Loening OL-5		V102
3	Loening OL-5		V103
4	Vought UO-4	404	V104
5	Vought UO-4	405	V105
8	FBA 17HT4		V107
9	Douglas O-38C		V108
10	Consolidated N4Y-1	310	V110
27	Douglas RD	227	V106
28	Douglas RD-1	128	V109
29	Douglas RD-2	129	V111
51	General Aviation PJ-2 (FLB51)	251	V116
52	General Aviation PJ-1 (FLB52)	252	V112
53	General Aviation PJ-1 (FLB53)	253	V113
54	General Aviation PJ-1 (FLB54)	254	V114
55	General Aviation PJ-1 (FLB55)	255	V115
	Douglas RD-4	130	V125
	Douglas RD-4	131	V126
	Douglas RD-4	132	V127
	Douglas RD-4	133	V128
	Douglas RD-4	134	V129
	Douglas RD-4	135	V130
	Douglas RD-4	136	V131
	Douglas RD-4	137	V132
	Douglas RD-4	138	V133
	Douglas RD-4	139	V134
	Grumman JF-2	161	V135
	Grumman JF-2	162	V136
	Grumman JF-2	163	V137
	Grumman JF-2	164	V138
	Grumman JF-2	165	V139
	Grumman JF-2	166	V140
	Grumman JF-2	167	V141
	Grumman JF-2	168	V142
	Grumman JF-2	169	V143
	Grumman JF-2	170	V144
	Grumman JF-2	171	V145
	Grumman JF-2	172	V146
	Grumman JF-2	173	V147
	Grumman JF-2	174	V148
	Vought O2U-2	301	V117
	Vought O2U-2	302	V118
	Vought O2U-2	303	V119
	Vought O2U-2	304	V120
	Vought O2U-2	305	V121
	Vought O2U-2	306	V122
	New Standard NT-2	311	V123
	New Standard NT-2	312	V124
	Stinson R3Q-1 (RQ-1)	381	V149
	Northrop RT-1	382	V150
	Lockheed R3O-1	383	V151

THE EARLY US COAST GUARD SERIAL SYSTEM INDIVIDUAL AIRCRAFT NAMES

The US Coast Guard in the thirties was able to make use of one of those fascinating peacetime luxuries that are normally limited to small organisations — the naming of individual aircraft. This practice was so common that many official Coast Guard communications, and nearly all press releases and newspaper stories, referred only to the name of the aircraft. These names appeared on each side of the nose and serve as an accurate means of identification of the individual aircraft. In photographs where the serial number is not visible, it is often the only means of positive identification.

Name	Type	3-Digit Serial
Acamar	General Aviation PJ-1	254
Acrux	General Aviation PJ-1	253
Adhara	Douglas RD-2	129
Aldebaran	Douglas RD-4	135
Alioth	Douglas RD-4	132
Altair	General Aviation PJ-1	252
Antares	General Aviation PJ-2	251
Arcturus	General Aviation PJ-1	255
Bellatrix	Douglas RD-4	138
Canopus	Douglas RD-4	139
Capella	Douglas RD-4	137
Deneb	Douglas RD-4	134
Mizar	Douglas RD-4	131
Procyon	Douglas RD	227
Rigel	Douglas RD-4	136
Sirius	Douglas RD-1	128
Spica	Douglas RD-4	130
Vega	Douglas RD-4	133

EARLY US COAST GUARD AVIATION SERIALS THE 'V' SERIES 1936-1945

Serial	Type	c/n	Period
V101	Loening OL-5		1926-1935
V102	Loening OL-5		
V103	Loening OL-5		
V104	Vought UO-4		1926-1934
V105	Vought UO-4		
V106	Douglas RD	703	1931-1939
V107	FBA 17HT4 (Schreck)		
V108	Douglas O-38C	1120	1931-1934
V109	Douglas RD-1	1000	1932-1939
V110	Consolidated N4Y-1		1932-1941
V111	Douglas RD-2	1122	1932-1937
V112	General Aviation PJ-1		1932-1941
V113	General Aviation PJ-1		
V114	General Aviation PJ-1		
V115	General Aviation PJ-1		
V116	General Aviation PJ-2		
V117	Vought O2U-2		1934-1940
V118	Vought O2U-2		
V119	Vought O2U-2		
V120	Vought O2U-2		
V121	Vought O2U-2		
V122	Vought O2U-2		
V123	New Standard NT-2		1934-1935
V124	New Standard NT-2		
V125	Douglas RD-4	1268	1934-1943
V126	Douglas RD-4	1269	
V127	Douglas RD-4	1270	
V128	Douglas RD-4	1271	
V129	Douglas RD-4	1272	
V130	Douglas RD-4	1273	
V131	Douglas RD-4	1274	
V132	Douglas RD-4	1275	
V133	Douglas RD-4	1276	
V134	Douglas RD-4	1277	
V135	Grumman JF-2	188	1934-1941
V136	Grumman JF-2	189	
V137	Grumman JF-2	190	
V138	Grumman JF-2	191	
V139	Grumman JF-2	192	
V140	Grumman JF-2	193	
V141	Grumman JF-2	194	
V142	Grumman JF-2	195	
V143	Grumman JF-2	196	
V144	Grumman JF-2	263	
V145	Grumman JF-2	264	
V146	Grumman JF-2	265	
V147	Grumman JF-2	266	
V148	Grumman JF-2	267	
V149	Stinson R3Q-1	74	1934-1941
V150	Northrop RT-1	74	1935-1940
V151	Lockheed R3O-1	1053	1936-1942
V152	OO-1 Viking		1931-1941
V153	OO-1 Viking		

EARLY US COAST GUARD AVIATION SERIALS THE 'V' SERIES 1936-1945

Serial	Type	c/n	Period
V154	OO-1 Viking		
V155	OO-1 Viking		
V156	OO-1 Viking		
V157	Waco J2W-1	4545	1937-1939
V158	Waco J2W-1	4546	
V159	Waco J2W-1	4547	
V160	Fairchild J2K-1		1937-1941
V161	Fairchild J2K-1		
V162	Fairchild J2K-2		1937-1941
V163	Fairchild J2K-2		
V164	Hall Aluminium PH-2		1938-1944
V165	Hall Aluminium PH-2		
V166	Hall Aluminium PH-2		
V167	Hall Aluminium PH-2		
V168	Hall Aluminium PH-2		
V169	Hall Aluminium PH-2		
V170	Hall Aluminium PH-2		
V171	Curtiss SOC-4 Seagull		1938-1941
V172	Curtiss SOC-4 Seagull		
V173	Curtiss SOC-4 Seagull		
V174	Grumman JRF-2 Goose	1063	1939-1948
V175	Grumman JRF-2 Goose	1064	
V176	Grumman JRF-2 Goose	1065	
V177	Hall Aluminium PH-3		1940-1944
V178	Hall Aluminium PH-3		
V179	Hall Aluminium PH-3		
V180	Hall Aluminium PH-3		
V181	Hall Aluminium PH-3		
V182	Hall Aluminium PH-3		
V183	Hall Aluminium PH-3		
V184	Grumman JRF-2 Goose	1076	1939-1948
V185	Grumman JRF-2 Goose	1077	
V186	Grumman JRF-2 Goose	1078	
V187	Grumman JRF-2 Goose	1079	
V188	Lockheed R5O-1	2008	1940-1946
V189	Consolidated PBY-5		1941-1943
V190	Grumman JRF-3 Goose	1085	1939-1948
V191	Grumman JRF-3 Goose	1086	
V192	Grumman JRF-3 Goose	1087	
V193	Naval Aircraft Factory N3N-3		1941-1945
V194	Naval Aircraft Factory N3N-3		
V195	Naval Aircraft Factory N3N-3		
V196	Naval Aircraft Factory N3N-3		
V197	Grumman J4F-1 Widgeon	1222	1941-1948
V198	Grumman J4F-1 Widgeon	1223	
V199	Grumman J4F-1 Widgeon	1224	
V200	Grumman J4F-1 Widgeon	1225	
V201	Grumman J4F-1 Widgeon	1226	
V202	Grumman J4F-1 Widgeon	1227	
V203	Grumman J4F-1 Widgeon	1228	
V204	Grumman J4F-1 Widgeon	1229	
V205	Grumman J4F-1 Widgeon	1253	
V206	Grumman J4F-1 Widgeon	1254	

EARLY US COAST GUARD AVIATION SERIALS THE 'V' SERIES 1936-1945

Serial	Type	c/n	Period
V207	Grumman J4F-1 Widgeon	1255	
V208	Grumman J4F-1 Widgeon	1256	
V209	Grumman J4F-1 Widgeon	1257	
V210	Grumman J4F-1 Widgeon	1258	
V211	Grumman J4F-1 Widgeon	1259	
V212	Grumman J4F-1 Widgeon	1260	
V213	Grumman J4F-1 Widgeon	1261	
V214	Grumman J4F-1 Widgeon	1262	
V215	Grumman J4F-1 Widgeon	1263	
V216	Grumman J4F-1 Widgeon	1264	
V217	Grumman J4F-1 Widgeon	1265	
V218	Grumman J4F-1 Widgeon	1266	
V219	Grumman J4F-1 Widgeon	1267	
V220	Grumman J4F-1 Widgeon	1268	
V221	Grumman J4F-2 Widgeon	1269	
V222	Vultee SNV-1 Valiant		1942-1945
V223	Vultee SNV-1 Valiant		
V224	Grumman JRF-5G Goose		1943-1954
V225	Grumman JRF-5G Goose		
V226	Grumman JRF-5G Goose		
V227	Grumman JRF-5G Goose		
V228	Grumman JRF-5G Goose		
V229	Grumman JRF-5G Goose		

US COAST GUARD AIRCRAFT SERIAL NUMBERS 1946-1991

Serial	Type	No.	Name
1230-1236	Sikorsky HO3S-1G	9	
1239	Kaman HK-225	1	Mixmaster
1240-1243	Grumman UF-1G	4	Albatross
1244-1251	Sikorsky HO5S-1G	8	
1252-1258	Sikorsky HO4S-1/2G	7	
1259-1267	Grumman UF-1G	9	Albatross
1268-1270	Bell HTL-3/5	3	
1271-1280	Grumman UF-1G	10	
1281	Sikorsky HO4S-3G	1	
1281-1283	Martin RM-1Z/VC-3A	2	
1284-1287	Martin P5M-1G	4	Marlin
1288-1294	Grumman UF-1G	7	Albatross
1295-1297	Martin P5M-1G	3	Marlin
1298-1310	Sikorsky HO4S-3	13	
1311	Grumman UF-1G	1	Albatross
1312	Martin P5M-2G	1	Marlin
1313-1317	Grumman UF-1G	5	Albatross
1318-1320	Martin P5M-2G	3	Marlin
1321-1322	— not used —		
1323-1331	Sikorsky HO4S-3G	9	
1332-1336	Sikorsky HUS-1G	5	
1337-1338	Bell HUL-1G	2	
1339-1342	Lockheed R8V-1G	4	Hercules
1343	Sikorsky HUS-1G	1	
1344-1351	Lockheed R8V-1G	8	Hercules
1352-1379	Sikorsky HU2S-1G	28	Seaguard
1380-1381	Grumman VC-4A	2	Gulfstream. 1381 cancelled
1382-1413	Sikorsky HH-52A	32	Seaguard
1414	Lockheed EC-130E	1	Hercules
1415-1429	Sikorsky HH-52A	15	Seaguard
1430-1438	Sikorsky HH-3F	9	Pelican
1439-1450	Sikorsky HH-52A	12	Seaguard
1451	— not used —		
1452-1454	Lockheed HC-130H	3	Hercules
1455-1466	Sikorsky HH-52A	12	Seaguard
1467-1497	Sikorsky HH-3F	31	Pelican
1498-1499	— not used —		
1500-1504	Lockheed HC-130H	5	Hercules
1600-1603	Lockheed HC-130H	4	Hercules
1700-1721	Lockheed HC-130H	22	Hercules
1790	Lockheed HC-130H	1	Hercules
2101-2141	Falcon HU-25A	41	Guardian
3501-3508	Grumman E-2C	8	Hawkeye
4104-4123	Aérospatiale HH-65A	20	Dolphin — not used
6001-6032	Sikorsky HH-60J	32	Jayhawk
6501-6596	Aérospatiale HH-65A	96	Dolphin
8101-8102	Schweizer RG-8A	2	Condor

Notes: 1790 allocated to a Hercules as it denotes the year the US Coast Guard was formed.

Serials for the first twenty HH-65A Dolphins were re-allocated.

CHAPTER FOUR
The Rotary Wing Period

Sikorsky Aircraft has been the prime manufacturer of helicopters for the US Coast Guard since 1943. This long and distinguished partnership has written into the annals of aviation one of the most inspiring chapters of how dedicated men and dependable machines can serve the most noble of all human pursuits which, as stated in the Coast Guard mandate, is 'to perform any and all acts to rescue and aid persons, and protect and save property.'

During 1941 the Coast Guard was seriously interested in developing the helicopter for search and rescue. Lieutenant Commander William Kossler had represented the USCG on an inter-agency board formed as early as 1938 for the evaluation of experimental aircraft, including the helicopter. However, World War II interrupted these plans.

Igor I Sikorsky, founder of the company, and designer and builder of the world's first practical helicopter, in 1943 wrote: 'In my mind, the helicopter's work as a rescue vehicle is one of the brightest and satisfying in aviation history. In this humane work, the helicopter has served mankind in thousands of emergencies where no other vehicle could have done the job.'

It is in this 'humane work' and in many other vital missions, that the US Coast Guard, using helicopters, has logged a record of achievement for which it, the nation and the world can be justifiably proud.

During 1939, an event took place which was soon to add new dimensions to the Coast Guard's mission effectiveness. The event was the first flight, on 14 September 1939, of Igor Sikorsky's VS-300 helicopter, which four years later led to the establishment of the helicopter industry and the world's first helicopter production line at Stratford, Connecticut.

The Coast Guard, then under the jurisdiction of the US Navy, was one of the first military services to recognise the potential of the helicopter, and over the past decades it has earned the credit for pioneering and perfecting many rotary wing applications.

In April 1942, after witnessing early flight tests of the Sikorsky XR-4, a test model which had been built for the US Army Air Forces, Commander W A Burton, Commanding Officer of the USCG air station at Brooklyn, New York, wrote in his report: 'The helicopter in its present stage of development has many of

L. TO R. - 1. E.H. FRAUENBERGER A.S.
2. J.A. BOONE A.M.M. 1C.
3. L. BRZYCKI A.C.M.M.
4. A.N. FISHER LT.
5. F.A. ERICKSON LT. COMDR.
6. O.M. HELGREN LT.
7. O.F. BERRY A.C.M.M.
8. W.J. WOODCOCK A.M.M. 1C.

The first USCG helicopter located at the Sikorsky factory at Bridgeport, Connecticut. It was headed by Lieutenant Commander F A Erickson, the first Coast Guardsman to qualify as a helicopter pilot, and who pioneered rotary wing development for military use. Photo taken on 7 July 1943. *(USCG)*

the advantages of the blimp and few of the disadvantages. It hovers and manouvers with more facility in rough air than the blimp. It can land and take off in less space. It does not require a large ground handling crew. It does not need a large hangar. There is sufficient range (about two hours) in this particular model to make its use entirely practical for harbor patrol and other Coast Guard duties.'

On the basis of this recommendation, Commander F A Leamy, Chief of Aviation Operations, Coast Guard headquarters, in June 1942, advocated the purchase of several Sikorsky helicopters for training and experimental development.

Lieutenant Commander Frank A Erickson, an officer destined to play a major role in the adaption of the helicopter to Coast Guard use, described in a memorandum in June 1942, a flight demonstration he had witnessed as a Sikorsky plant in Stratford, Connecticut. He noted that 'the life saving and law enforcement possibilities of the helicopter have heretofore been especially stressed. However, this machine can fulfill an even more important role, that is in providing aerial protection for convoys against submarine action, an important function of Coast Guard Aviation.'

Commander Erickson suggested that five of the Sikorsky R-4s,

Floyd Bennett Field, Brooklyn, New York, on 21 December 1943 with a Sikorsky HNS-1 helicopter BuNo 46445 of the USCG seen hovering with a stretcher suspended below. The Coast Guard initiated many 'firsts' with this early helicopter under the guidance of Commander Frank A Erickson, USCG pioneer helicopter pilot. *(USCG)*

then being built for the US Army Air Force, and which would be in full production by early 1943, could give far greater protection for a convoy thán a similar number of blimps or aircraft. This type of helicopter was reputed to be able to carry a crew of two, a 325lb depth charge, radio and other equipment, and fuel for about four hours.

The proposal was endorsed by Commander Burton, who pointed out that this was an excellent opportunity for the Coast Guard with a very modest appropriation, to initiate and proceed with the development of the helicopter.

During January 1943, the United Kingdom placed an order for 200 Sikorsky helicopters and opened negotiations for 800 more. At this time, the Coast Guard had two Sikorsky XR-6 helicopters on order. Shortly after this, Vice Admiral R R Waesche, USCG Commandant, held a conference with Admiral E J King, Chief of Naval Operations, concerning the lagging development of helicopters for naval service. On 15 February 1943 Admiral King issued a directive to the Chief of the Bureau of Aeronautics that effectively launched a development programme. He ordered that the testing and evaluation of helicopters be initiated and carried through quickly to determine their practical value for operating from merchant ships in ocean convoys. The tests were to be conducted by the Bureau of Aeronautics and the Coast Guard, with the US Maritime Commission providing a merchant ship with a suitable platform, the US Army Air Force providing three Sikorsky YR-4A helicopters and the Coast Guard providing the aviators.

A 'Combined Board for the Evaluation of the Helicopter in Anti-Submarine Warfare' was formed consisting of representatives of the Bureau of Aeronautics, Great Britain, the Coast Guard and the Commander-in-Chief, US Fleet.

FIRST TESTS

The first tests got underway on 7 May 1943. Fifty observers were on hand when Colonel Frank Gregory, USAAF, gave a convincing demonstration of the Sikorsky R-4 aboard the USS *Bunker Hill*. He made more than twenty flights from a small strip deck previously used for deck cargo. The helicopter successfully completed every landing with great precision.

Enthusiasm ran high following these first tests. Action was taken to provide two ships with landing decks for helicopters. They were the *Governor Cobb*, a merchant vessel, to be operated by the Coast Guard, and the *Daghestan*, also a merchant vessel being operated by the British under Lend-lease.

The US Navy accepted its first Sikorsky HNS-1 helicopter on 16 October 1943 with the second and third following on 2 November 1943. Events moved swiftly during November, and by the 19th, the USCG air station at Floyd Bennett Field, Brooklyn, New York, was officially designated as a helicopter training base equipped with three Sikorsky HNS-1 helicopters. The open sea trials on board the *Daghestan* took place commencing 29 November off Bridgeport, Connecticut, with a total of 328 landings made by both British and United States pilots.

Meanwhile at the Brooklyn air station, a special movable platform forty feet by sixty feet had been developed that

The Sikorsky HOS-1 helicopter was manufactured by Nash Kelvinator, some twenty-seven operating with the USCG between 1945/49. Depicted is BuNo 75623 from the second batch. The type was involved in shipboard trials with the HNS-1 on the *Governor Cobb* with the helicopters operating from Brooklyn, New York. *(USCG)*

could simulate the motion of the ship's deck at sea. Further development work on stretchers, slings and related equipment progressed rapidly.

One of the many people involved in this work was Igor Sikorsky's son, Sergei, who was serving in the Coast Guard at this time. He was the crew chief on one of the original HNS-1s and spent nearly two years working with other military personnel on the development and perfection of the helicopter rescue hoist. The first actual use of the hydraulic winch hoist in a rescue occurred on 29 November 1945 during a severe storm in the vicinity of Bridgeport, Connecticut. Two crew members of a barge which had broken away from its tow and was in imminent danger of breaking up were successfully rescued by means of a hoist-equipped Sikorsky helicopter sent out from the factory.

It was a tentative beginning. But the awkward, flimsy vertical lift vehicle of the 1940s showed the promise of great accomplishments to come. And the faith and persistance of these men who evaluated its initial trials were to open new horizons in man's conquest of the air.

ADVANCED TECHNOLOGY

With World War II over, the US Coast Guard was released from the US Navy to once more become a separate agency. This commenced an era that was to witness the Coast Guard's increasing reliance on the helicopter for many of its key maritime responsibilities. The helicopters it was to select in the years ahead were mostly designed, developed and manufactured by Sikorsky, each one a step forward in the evolutionary process of incorporating the latest features of advanced and proven technology.

Greater speeds, extended range, increased payload, improved rescue devices, amphibian on-the-water landing capability, and all-weather round-the-clock operational suitability — all became reality in rotary wing flight. And every one of these advances in a succession of helicopters enabled the dedicated and daring Coast Guard aviators to broaden their skills in performing their myriad tasks.

Commander Frank A Erickson became Commanding Officer of the USCG air station at Floyd Bennett Field, New York. He was the first USCG helicopter pilot and pioneered the development of rotary wing aircraft for practical military use on patrols and rescues. The USCG became responsible for training all Allied helicopter pilots at Brooklyn. A total of twenty-one Sikorsky HNS-1 helicopters were purchased by the Coast Guard at a unit cost of $43,940 and they served the US Coast Guard between 1943 and 1948.

During 1945 the Coast Guard purchased twenty-seven R-6A helicopters from Sikorsky, designated HOS-1G, these being built by Nash Kelvinator under licence from Sikorsky. First acceptance of this type was during September 1944 when the first was delivered to Floyd Bennett Field for USCG evaluation. It was powered by a Franklin 0-405-9 240hp engine and had a range of 245 miles and could carry three people. It had a cruise speed of 75 knots, and had a gross weight of 2900 pounds.

Unique photo taken at the US Naval Air Station Anacostia, Maryland, during 1949, depicting a USCG HO3S-1 '232' fitted with 'doughnut' type flotation gear. A total of nine operated with the service between 1947 and 1955, one being preserved by the USCG and now in the Naval Aviation Museum at Pensacola, Florida. *(Fahey via William T Larkins)*

The post-World War II years brought an explosion in the number of recreational boats and created a new search and rescue clientele. The helicopter was ideally suited to this mission. Able to react swiftly, it could lift entire pleasure boat crews from disaster, or in less trying circumstances, deliver dewatering pumps and fuel. Admittedly during its early years the helicopter had a major handicap — the pilot needed three hands in order to fly it. Soon, Coast Guard helicopters rescuing boaters in distress became a commonplace event.

Deliveries of a new Sikorsky helicopter to the US Navy commenced in November 1946, this being designated HO2S-1. The Coast Guard purchased two only at a unit cost of $86,000 in January that year. It had a cruise speed of 80 knots, a range of 230 miles and was powered by a Pratt & Whitney Wasp Jr R-985-AN5 engine. After initial evaluation the HO2S-1 was replaced by the HO3S-1G of which nine were purchased by the USCG at a unit cost of $91,977. Powered by a Pratt & Whitney

Only three Piasecki HRP-1 'Flying Banana' helicopters appeared on the USCG inventory, two being delivered in November 1948, and one the following month, this — BuNo 111826 — crashing in April 1951. Photo depicts a Coast Guard HRP-1 during winching exercises involving a litter patient. *(USCG)*

R-985-AW-5 engine, it had a cruise speed of 80 knots, an increased range of 240 nautical miles and a useful load of 1450lb. These were the first Coast Guard helicopters to be equipped with a rescue hoist, and some USCG models were fitted with inflatable bags which fitted over the wheels.

During the 1950s many versions of the Sikorsky S-55 were used by the US armed services including the Coast Guard. They played an important part in developing helicopter roles and techniques. Of classic helicopter configuration, with a single main rotor and anti-torque tail rotor, this was the first of the Sikorsky helicopters with adequate cabin space and lifting ability to permit satisfactory operation in the air/sea rescue role. Two versions were procured by the Coast Guard — HO4S-2G and HO4S-3G — the major difference being the engine. The unit cost was $177,530 and twenty-one were purchased while eight HRS-1s were borrowed from the US Navy and operated for several years. The type served the USCG from 1951 until 1966. It had a cruising speed of 79 knots, a range of 360 nautical miles and a useful load of 2250 pounds.

The versatility of the helicopter was demonstrated during a series of floods which occurred in the United States in the 1950s. To carry out this kind of rescue work, the helicopter had to hover among trees, telephone poles, television antennas and the like. In 1955, Coast Guard helicopters rescued more than 300 people as rivers overflowed in Connecticut and Massachusetts. In December of that year the USCG on-scene commander directed the rescue of thousands in California. Included among the fleet of twenty-one rescue aircraft were Coast Guard helicopters. In one incident a Sikorsky HO4S rescued 138 people during a twelve-hour period, this being accomplished by two air crews. The helicopter soon developed from a thoroughbred requiring a great deal of pampering to keep it flying, to a reliable workhorse. Rescues involved using the hydraulic hoist and the Coast Guard-designated rescue basket. All of the Coast Guard HO4S helicopters were fitted with a 'Tugbird', another USCG design which provided the unique capability of towing boats and ships of up to 794 tons.

The first United States helicopter to have metal rotor blades was the Sikorsky S-52 developed as a two-seater and powered by a 178hp Franklin engine, making its first flight on 14 February 1947. Eight of these helicopters were procured by the Coast Guard as HO5S-1G at a unit cost of $82,928. Top speed was 90 knots but unfortunately its size, short range and low life capability limited its effectiveness. It served the USCG between 1952 and 1955.

During 1947 the Coast Guard procured two four-seat Bell Model 47J helicopters, designated HTL-1 by the US Navy and HUL-1G by the USCG. These were used for survey of the New York Harbor area under the direction of the Captain of the Port of New York. They usually operated with floats attached, but could

A US Coast Guard Bell HTL-1 helicopter seen on 30 October 1953 lowering a line to a waiting Coast Guard lifeboat during a towing drill. The HTL-1 helicopter came from the Floyd Bennett Field CGAS located at Brooklyn, New York. *(USCG)*

be flown with skids. Smuggling, harbour pollution, sabotage and other maritime derelictions were all part of the task.

At a unit cost of $256,912 three Piasecki HRP-1 helicopters were purchased by the Coast Guard in 1948. Powered by a Pratt & Whitney R-1840-AN1 engine, they had a cruise speed of 64 knots and a range of 140 miles. Fitted with a rescue hoist, the HRP-1 carried a crew of two and could pick up eight survivors. They served until 1952 and were based at Elizabeth City air station in North Carolina. Also operated at this station was a single Kaman K-225 Mixmaster helicopter in the 1950s which the Coast Guard purchased for $37,684. It was powered by a Lycoming 0-435-A2 engine, cruised at 60 knots and had a range of 145 miles.

This Bell HTL-1 ex USAAF YH-13, helicopter retains its US Navy BuNo 122461 and has limited US Coast Guard identity markings and 'MISSIONS' inscribed on top of the fuselage. Two of the type were operated by the service between 1947 and 1955. *(USCG)*

After experience with the Bell HUL-1G helicopter, the Coast Guard purchased three Bell HTL-5s during 1952. Thirty-six had been procured by the US Navy powered by a Franklin 0-335-5 engine. Unit cost was $49,290 and the USCG used them on a variety of tasks, but again their small size and short range limited their effectiveness. Two more Bell HUL-1Gs, redesignated HH-13Q in 1962, were acquired by the Coast Guard in 1959 operating until 8 December 1967. They were based at Kodiak air station in Alaska and fitted with floats were normally used for ice reconnaissance from USCG cutters in the Bering Sea and the Gulf of Alaska. When operated from land the HUL-1G was fitted

Records reveal that this unique USCG helicopter, Kaman K-225 'CG-239' was one of three produced under a US Navy contract for evaluation of flying qualities. It was the seventh K-225 produced and had the US Navy BuNo 51917 prior to going to the Coast Guard. *(USCG)*

with skids. It was a four-place helicopter featuring a 400lb hydraulic hoist for rescue work, and was fully instrumentated for night flying. The large transparent plastic bubble at the front of the cabin gave the pilot a wide range of view. Photographs indicate that at least one Bell HTL-7 helicopter served with the Coast Guard, being based in Alaska.

Meanwhile during 1959 the Coast Guard purchased six new Sikorsky helicopters to replace the less capable HO4S types. This was the Sikorsky S-58 model designated HUS-1G which was a major step forward in payload. It had a gross weight of 13,000lb, a cruise speed of 84 knots, a useful load of 4725lb, and

A Sikorsky HO4S-3 '1331' of the USCG, piloted by Lt Benjamin Weems USCG, hovers over the site of a new heliport in New Orleans, Louisiana, on 3 September 1959. The helicopter was from the local USCG Air Detachment and the type was in production from 1949 to 1961 with 1281 being built. *(USCG)*

Scene at USCG Air Detachment, New Orleans, Louisiana, on 5 October 1959 are two Sikorsky HO4S-2G helicopters are replaced by two new HUS-1G machines. The older type flew 510 SAR missions, 53,943 miles representing 1023 hours flown over the previous year. After rework the HO4S-helicopters went back into service at Traverse City, USCG Air Station, Michigan. *(USCG)*

a range of 215 nautical miles. Hundreds of this model flew thousands of hours for many military air arms including the US Navy and the US Air Force. Internal accommodation included side-by-side seating for the two pilots and room for ten passengers. It could carry a droppable external fuel tank increasing the range to 500 miles.

TURBINE POWER

During 1963, the Coast Guard took delivery of its first turbine-powered helicopter, a version of the commercial model Sikorsky S-62, designated HH-62A and given the name Seaguard. First deliveries commenced in January 1963, and by early 1967 orders from the USCG totalled fifty-eight and total orders numbered ninety-seven. The HH-52A was originally designated HU2S-1G, but under a Department of Defense directive dated 6 July 1962 the designations allocated to the US Navy and US Coast Guard were standardised with those of the US Air Force. This new helicopter workhorse was not only the first truly amphibious helicopter, but the Coast Guard's long standing requirement for a flying life boat. It had a watertight hull and outrigger sponsons which enabled it to alight on the water to retrieve victims from the sea. It also had a special rescue platform which folded out and down over the water. With a speed of 85 knots, a gross weight of 8100lb and a range of 412 nautical miles, it soon came to be called the flying life boat. The helicopter was powered by a T58-GE-8B turbine engine, and in addition to shore-based duties it was deployed on USCG icebreakers and cutters for extended periods. The purchase price was quoted as ranging from approximately $250,000 to over $500,000, just for the airframe.

Next to join the Coast Guard fleet of helicopters, in 1969, was the large twin-turbine, amphibious Sikorsky S-61 model, designated HH-3F and named Pelican with the force. Forty of these helicopters were delivered over the next four years. Built by Sikorsky to Coast Guard specifications, it has two 1500hp gas turbines, a top speed of 157 mph and a range of 700 miles. Gross weight is 22,050lb, it carries a useful load of 8537lb and

Lieutenant R L Cook, USCG, about to board a Bell HUL-1G helicopter '1338' at Kodiak, Alaska, on 15 May 1961. Two of the type were used by the service for the first time during the summer of 1959 on the Bering Sea Patrol. They operated from the USCG icebreaker *Northwind*. *(USCG)*

Sikorsky HH-3F Pelican '1468' from Brooklyn USCG air station, New York, seen on a flight over the Atlantic seaboard. It carries a crew of four — pilot, co-pilot, navigator and crew-chief, with seats for six passengers. It has provision for twenty survivors or nine litters. It is similar to the Sikorsky CH-3E. *(USCG)*

the last one was delivered in 1973. Price for the basic airframe only was approximately $900,000.

The HH-3F is designed to provide extended range for search and rescue (SAR) operations over the open sea. It has the speed, range, and capacity to carry sophisticated electronics gear for automatic navigation, communications and radar necessary for the mission. There are seats for six passengers and provisions for nine stretchers, plus storage for all rescue equipment. With the aircraft on the water, a special rescue platform can be extended from the cabin door for safe retrieval of disabled survivors. The increased fuel capacity allows for the long range required for the SAR mission. Alternate seating provisions are available for twenty passengers. Its multi-mission capabilities make the Pelican a cost-effective solution for the SAR utility requirements of many nations.

Twin-turbine engine propulsion combined with an integrated flight system, using heading and attitude sensors and radio navigational aids, provides a complete all-weather IFR instrument capability. Weather search airborne radar equipment further improves the HH-3F's search performance.

When the Coast Guard took delivery of the fortieth and final Sikorsky HH-3F helicopter on 10 July 1973, Rear Admiral William A Jenkins, then Chief of the Office of Operations, cited the long USCG-Sikorsky relationship going back as far as 1930 and the

The US Navy loaned the USCG two Bell TH-13N helicopters which still carried the pre-1962 designation HTL-7. They carried BuNos 145848 and 145853 and were loaned during 1965 and had a value of $163,696 each. Depicted is BuNo 145848 HTL-7 seen hangared on a US Coast Guard vessel. Date and location unknown. *(USCG)*

S-39 fixed-wing amphibian. He referred to the HH-3F as the most sophisticated search and rescue helicopter in operation and that it was a valuable tool with a performance level far in excess of what had been anticipated.

During April 1978, the Coast Guard announced a request for bid proposals for the design and construction of ninety new short-range recovery (SRR) helicopters. Designed as a replacement for the HH-52A Seaguard, Sikorsky submitted initially a proposal for its twin turbine-powered S-76 with the belief that it would meet the requirements set down by the Coast Guard. Later Sikorsky withdrew the S-76 from the new SRR requirement.

It was during 1977 the USCG issued an acquisition programme to give the service a new short-range recovery helicopter by late 1979 or early 1980. Proposal response from manufacturers was required by December 1977 — it was issued in September — with a Coast Guard decision planned for mid-August 1978. The short programme cycle was due to the USCG, in consultation with the US Navy, in requiring all candidates to be able to receive a Federal Aviation Administration (FAA) derivative type certificate or a military equivalent in the acquisition time frame.

The requirements for the USCG prime mission, Search &

The versatile Sikorsky HH-52A Seaguard amphibious rescue helicopter is seen landing on a heliport on top of the commercial oil rig *Movible No.3* position in the Gulf of Mexico off the coast of Louisiana on 16 October 1963. *(USCG)*

Rescue, stated that the short-range recovery candidate must take-off, cruise outbound at 1000ft in excess of 100 knots, travel 150 nautical miles from base, hover for thirty minutes and then pick up three 170lb survivors, return to base and have ten per cent of the available fuel, or twenty minutes of fuel remaining, whichever is the greater, as reserve fuel at shutdown.

Helicopter manufacturers who competed for the Coast Guard request requirements, in addition to Sikorsky with the S-76, included Bell Helicopter Textron with a utility version of its Model 222, and Aérospatiale with a modified version of its SA 365. The Sikorsky S-76 was reputed to be able to pick up six 200lb survivors and satisfy the Coast Guard rescue profile requirements.

Aérospatiale conducted studies to see if the SA 365 would be viable as a candidate for the programme. It was estimated it would have to install a pressure refuelling system and adapt the nose to accommodate a radar. The main change to the SA 365 was to be replacement of the two Turboméca Arriel engines to conform with the Buy American Act. A version of the helicopter had flown with US-manufactured Lycoming LTS 101 engines. They proposed to enrol US vendors for the avionics, auxiliary power unit (APU) and the environmental controls. Aérospatiale emphasised the single pilot instrument flight rules ability, the hover stability and the fan-in-fin as attributes to the SA 365.

HH-65A DOLPHIN

On 13 November 1984 in a news release the US Department of Transportation announced that the US Coast Guard had announced plans for conditional acceptance of HH-65A Dolphin helicopters from the Aérospatiale Helicopter Corporation. The first of ninety-six helicopters ordered was to be delivered from the Aérospatiale assembly plant in Grant Prairie, Texas. Acceptance of the aircraft had been delayed by snow ingestion problems. The twin-engine Dolphin has a range of 400 nautical miles, and replaced the HH-52A Seaguard as the USCG's short-range helicopter. It can be carried on board the USCG medium and high endurance cutters, and is being used in law enforcement and ice surveillance as well as Search & Rescue.

The US Coast Guard, over its seventy-five years of aviation history since 1916, had relied almost wholly for its aircraft upon types proved and selected with the US Navy, and in odd cases with the US Air Force. A major break occurred when the requirement for the new SRR helicopter was fulfilled by selection of a version of the French-built SA 365. The USCG version was identified as the SA 365G receiving the Department of Defense designation HH-65A. First flight was made in France on 23 July 1980, this aircraft being shipped to the United States to serve as one of the two trials aircraft for the HH-65A. Introduction into service was delayed by more than two years by a combination of problems, one of which was the susceptibility of the engines to snow ingestion.

Deliveries began on 19 November 1984 with the first six helicopters going to the Coast Guard Training Center located at

After delivery in January 1987, this HH-65A '6544' Dolphin was assigned to CGAS Savannah, Georgia, the following month. It was transferred to Port Angeles CGAS, Washington, in July 1988. The helicopter was designed for short-range SAR operations for the USCG and is fully instrumentated for day and night flights. *(Ralph Peterson)*

Mobile, Alabama. The last HH-65A was delivered on 24 April 1990. However, Lycoming LTS-101 engine performance and reliability problems have prompted the Coast Guard to plan flight tests of a re-engined Allison/Garrett T800-powered HH-65A helicopter during 1991. The LTS-101 engine has had a troubled operational history since it began service in 1984. It has experienced numerous in-flight shut downs and has suffered from power turbine wheel cracking problems. In addition the engine has operated at higher temperatures than initially designed. These problems have resulted in the engine, which was supposed to accumulate about 2400 hours between overhauls, lasting less than 600 hours.

Early in 1990 the Commandant of the US Coast Guard, Admiral Paul Yost, said, 'I have made a rule of thumb as the Commandant that I will never again buy a helicopter or an airplane that was not a DoD-supported piece of equipment.'

SIKORSKY HH-60J JAYHAWK

On 30 September 1986 a news release from United Technologies Sikorsky Aircraft at Stratford, Connecticut, revealed that the Department of the US Navy had announced the awarding of a $84.5 million contract to Sikorsky Aircraft for an initial production of five Combat Search & Rescue Special Warfare Support (HCS) helicopters for the US Navy, and two Medium Range Recovery (MRR) helicopters for the US Coast Guard.

The Sikorsky HH-60J Jayhawk will replace the Sikorsky-built HH-3F Pelican helicopter in the MRR mission. In its search and rescue (SAR) role, the new helicopter can fly up to 300 miles offshore and maintain an on-scene endurance for forty-five minutes. With a crew of four, the HH-60J can transport at least six survivors from the maximum SAR radius. Other missions will include drug interdiction, environmental protection, offshore enforcement of laws and treaties, aids to navigation and logistics support and other missions that may be required by joint Coast Guard — US Navy maintenance of the US Maritime Defense Zone.

The MRR aircraft is compatible with the helicopter decks of the 378ft Hamilton class and the 270ft Bear class Coast Guard cutters. Requirements also call for the helicopter to perform in violent storm-force winds up to Beaufort Scale 11, and over exceptionally heavy sea states. With its proven airframe derived from the extensive US Navy Sikorsky SH-60 family, its extended range and advanced avionics and communications suites, including a search-weather radar, the new Jayhawk offers a dramatic upgrade in ship-helicopter team capability in drug interdiction missions. Delivery of the first of twenty-four HH-60J helicopters on contract was scheduled for March 1990 to be delivered to the US Navy Test Center at Patuxent River, Maryland, for test and final evaluation. The Coast Guard has stated a need for thirty-two of the new helicopters. Cost is approximately $12 million each. The first air station to re-equip with the type will be Mobile, Alabama, for training, followed by Cape Cod, Massachusetts, and San Francisco, California. Final delivery is planned for 1993.

Sikorsky rolled out the first HH-60J on 14 September 1989 in a ceremony that coincided fifty years to the day with the first successful helicopter flight of aviation pioneer Igor I Sikorsky in the VS-300.

A total of six Sikorsky CH-3E Sea King helicopters have been acquired by the USCG from the US Air Force. Three have been modified for SAR missions with auxiliary fuel tank for increased range, a Loran 'C' navigation system plus a AN/APN-215 radar. Three will remain in storage at the Elizabeth City air station as replacement for spare parts.

During more than forty-five years of association, a dozen Sikorsky helicopter types have served with the US Coast Guard, the history records revealing 254 aircraft in all. The service, using the Sikorsky R-4/HNS-1 helicopter, defined SAR techniques and developed the rescue hoist during World War II at its air station at Brooklyn, New York. According to records the US Coast Guard with its helicopters can claim to have saved over 56,000 lives.

CHAPTER FIVE
Coast Guard Aircraft Acquisition

After the cessation of hostilities and the armistice after World War I, the US Coast Guard was returned to the US Treasury Department from the US Navy. The former Naval Air Station at Morehead City, North Carolina, was made available as a temporary Coast Guard air station. The US Navy loaned the Coast Guard six surplus Curtiss HS-2L flying boats. Due to lack of funds this air station was discontinued after only a year and the aircraft returned to the US Navy.

Coast Guard Aviation lay dormant through the years 1922, 1923 and 1924. In 1925 Lieutenant Commander C C von Paulsen, on his own initiative, borrowed a single Vought UO-1 seaplane from the US Navy for the period of one year. He purchased a

Depicted is a model of the Grumman G-3, an amphibian which was proposed to the USCG in 1930, but never built. It was a twin-engined parasol monoplane flying-boat. Grumman, like Sikorsky with its wide range of USCG helicopters provided the service with many of its operational tools. *(Grumman)*

surplus US Army tent for a single dollar. The US Department of Fisheries owned a small island in Gloucester Harbor, Massachusetts, and allowed von Paulsen to set up a temporary base there. This was the famous Coast Guard 'Ten Pound Island' air base. Again, due to lack of funds, the base was discontinued a year later, and the Vought returned to the US Navy.

It was the Coast Guard's success with the UO-1 that finally prompted Congress to appropriate the money which resulted in the procurement of the first five aircraft which the service actually owned. These were assigned to the air stations at Gloucester and the newly established Cape May USCG air base in New Jersey. Both these stations were active in rescue operations and anti-smuggling patrols during the period 1926-1928.

Despite continuing problems with funding Coast Guard Aviation continued to grow. Until 1932, funds were so lacking that the service could not usually afford to establish aircraft specifications. Instead it was forced to utilise US Army, US Navy or even civilian designs. A modest additional appropriation in 1931 enabled the service for the first time to acquire aircraft which were suited to its specific needs. It was the urgent requirement to stop liquor smuggling that finally motivated Congress to begin making available the funds necessary to fully support the Coast Guard's expanding aviation fleet.

On 9 March 1934 the Secretary of the Treasury directed that all aviation activities of the Treasury Department be consolidated under the Coast Guard. Overnight fifteen Customs Service aircraft, confiscated from rum runners and other offenders, were transferred to the Coast Guard. This increase in the aviation fleet was short-lived after several crashed and all but two were eventually condemned in the interest of safety and standardisation. In addition six Vought O2U-2s were transferred from the US Navy during 1934.

It is not generally known that a number of aircraft were borrowed over the years, mainly from the US Navy. During 1931, the year that the Coast Guard contract for the General Aviation Flying Life Boat was awarded, the Coast Guard borrowed a Naval Aircraft Factory PN-12 patrol flying boat for experimental research in connection with a further proposed design for a new Coast Guard aircraft.

In 1937 the Coast Guard Engine School & Repair Base at Norfolk, Virginia, received a single Martin T4M-1 torpedo-bomber scout biplane from nearby Naval Air Station Norfolk. Powered by a Pratt & Whitney engine it was used for the detection of engine problems and for ground instruction. This aircraft did not carry a USCG serial as it was not operational on the USCG inventory.

On 8 July 1940 two Great Lakes BG-1 carrier-based dive bombers went to the Engine School & Repair Base at Norfolk. They had served with the US Navy since 1935 and transferred

Speed is essential in all rescue operations. During and after World War II a handful of US Navy types were evaluated by the USCG in an effort to improve their efficient air sea rescue service. Types used included two Curtiss SB2C-4 Helldivers equipped with rafts and ration packs. Depicted is a Curtiss SC-1 Seahawk scout seaplane also evaluated, which, although a single-seater could carry a patient on a litter in the fuselage. *(USCG)*

from the nearby US Navy establishment. They had the paperwork reference 'NAS Norfolk letr. NA8/L11-6/P21(40-ly) of 7-8-40'. The BG-1 with US Navy BuNo 9506 had completed 1433 hours of flight, and BuNo 9519 1161 hours. Both were powered by a 750hp Pratt & Whitney R-1535-66 engine. No US Coast Guard serial numbers were allocated.

Having entered the 1930s with but five aircraft, the Coast Guard was to end that decade with ten times more, though it hardly represented a moderate force when one contemplated the far-reaching responsibilities involved. The service was standing on the threshold of a great expansion necessitated by its role in World War II. Early pioneers of aviation history would have been more than amazed at the forecast changes. Unlike its previous wartime affiliation under the US Navy, this smallest service was ready to flex its own musciles. The US Navy would be in control, but this time the Coast Guard warriors would retain their own identity for the forthcoming conflict. Restricted to no more than a couple of dozen aircraft during any previous year in its existance, suddenly the service was to experience an abundance of types. Admittedly compared to US Army and US Navy activities it was not much, but to the Coast Guard aviators it seemed like a veritable fleet — as well as a challenge.

PROCUREMENT IN REVERSE

A US Coast Guard contract in 1935 for six additional Grumman JF-2s was delivered during 1936 with the sixth aircraft '175' being traded to the US Navy. It was delivered from the

Grumman factory on 26 November 1935 to the US Marine Corps air station at Quantico, Virginia. In return the US Coast Guard received the Lockheed XR30-1 Electra.

During 1941, four Grumman JF-2s from the Coast Guard — V135, V141, V144 and V146, were traded to the US Navy in exchange for four Naval Aircraft Factory N3N-3 trainers. They were needed to supplement the increasing pilot training programme of the Coast Guard. These aircraft retained their US Navy primary training colours of yellow overall, but did have the US Coast Guard vertical tail stripes on the rudder, and the name and serial on the side of the fuselage.

While current procurement practices are more sophisticated than those of the past, the US Coast Guard continues to be, because of its small size, still subject to strict funding limitation. During the 1960s the Coast Guard's Planning, Programming and Budgeting System (PPBS) was instituted and is the mechanism used to ensure the maximum benefit from the funding available.

Referring back to 1915, it was a Curtiss F boat in which the two pioneers of Coast Guard aviation, 2nd Lt Norman B Hall and 3rd Lt Elmer F Stone from the USCG cutter *Onondaga* learned to fly at the Curtiss Aeroplane & Motor Company School located at Newport News, Virginia. One of the first successful flying boats developed, it proved the practicability of overwater search by aircraft. In World War I the Curtiss R6 observation and scout seaplane was flown by USCG aviators based on the cruiser USS *Huntington* and used by the fledgling air arm of the rescue service developed by the Coast Guard after hostilities ceased.

The prototype Pratt-LePage XR-1A 42-6581 was powered by two 450 hp R-985-AW-1 engines, driving two three-bladed rotors mounted on lateral out riggers. It carried a crew of two in tandem. This machine provided the basic idea for the series of McDonnell projects of 1944. The USCG suggested using the wings from the RD-4 Dolphin, and a set of XR-1 rotors fitted. *(USAF Museum via David E Menard)*

During World War II, the Coast Guard again operated US Navy aircraft, most being eventually turned back to the Navy. Some were scattered among the Mutual Defense Assistance Programme (MDAP), US Air Force, War Assetts Administration, the US Fish & Wild Life Service, and the National Advisory Committee for Aeronautics (NACA). A few were retained by the Coast Guard, to be eventually replaced by more modern aircraft.

During 1945 a single North Amerian B-25J Mitchell 431357 apparently was unofficially borrowed from the US Army Air Force receiving the 'US COAST GUARD' inscription on the fuselage. Also during 1945, the Coast Guard acquired two US Navy Curtiss SB2C-4 Helldiver scout bombers. They were involved in research and flight test into high-speed search and rescue by dropping rafts and rations to survivors while awaiting a rescue by a surface vessel. Powered by a Wright R-2600-8 engine, they had a speed close to 230 knots. Both Helldivers were located at San Diego air station, California, where they were also used for pilot training and served until 1947.

RESEARCH & DEVELOPMENT

Intent on continuing its superiority in aerial surveillance, the US Coast Guard's Office of Research & Development (R&D) is always actively investigating the use of modern state-of-the-art equipment to improve its capabilities. Over the past ten years or

An artist's impression of the ultimate 'Flying Life Boat' design for the USCG. It would be capable of being towed and dropped in the ocean and taxi-ing under its own power to pick-up survivors. Apparently many such designs and drawings originated from 'doodles' on an engineer's desk in the immediate post-war years. *(USCG)*

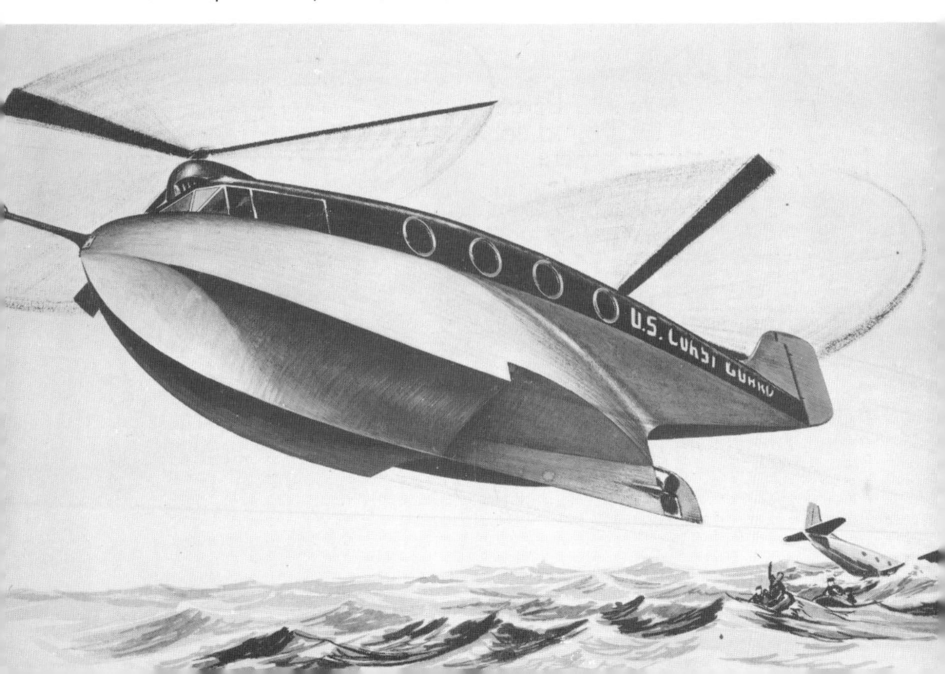

This glider-borne lifeboat design for the USCG was initiated in 1945, and developed by the service and the Stevens Institute of Technology. It was eventually abandoned in 1947. Commander Richard D Schmidtman USCG, later Rear Admiral, was the project engineering officer. *(USCG)*

more, this department has been responsible for the evaluation and introduction of major projects such as the development of an all-weather, day/night, multisensor system for the HU-25A Guardian, including the HU-25B and the modified HU-25C Interceptor medium-range search aircraft. Development of a forward-looking infra-red (FLIR) sensor for the HH-65A Dolphin helicopter, and looking ahead, investigation of a satellite-aided Search & Rescue system, are the responsibility of this department. Together, these systems have the potential for dramatic improvement in locating vessels in distress and other tasks.

Under the direction of the Chief, Aviation Division, and based at US Coast Guard headquarters in Washington, DC, a Facility Manager is appointed for each new aircraft type working closely with the manufacturer. In some cases an Aircraft Programme Office (APO) is located at the factory, an example being the APO set up at the Lockheed Aircraft Company in Marietta, Georgia, which provided integrated logistic support development for future Coast Guard procurement of the ubiquitous HC-130H Hercules. The APO was closed in June 1988 after the last Coast Guard-procured HC-130H was delivered. However, now a USCG Hercules has been delegated to be modified to carry the twenty-four foot diameter rotating radar dome on top of the fuselage. This is the AN/APS-125 radar as fitted to the USCG Grumman E-2C Hawkeye airborne early warning picket.

Management responsibility for the Coast Guard rests with the Commandant. He carries out this responsibility through the Chief of Staff, who co-ordinates the efforts of the Programme Directors. The responsibilities of the Commandant and Chief of Staff involve the overall objectives of the USCG while those of the Programme Directors involve the component parts, called Programmes, which are the means of achieving these objectives.

The USCG has many programmes, each representing a specific operating mission area, such as Search & Rescue, with a far as possible look ahead to future acquisitions.

The initial document in the planning process is the Commandant's Long Range View, where the Commandant sets forth his view of the environment in which the Coast Guard will be operating over the next twenty-five years. Based on these projections, specific policy guidance is provided. The Long Range View is a policy document which provides a common foundation for all planning at USCG headquarters and in the field.

On 9 January 1963 the USCG Commandant, Admiral E J Roland, and his acceptance committee attended the acceptance of the first HH-52A '1352' at the Sikorsky plant at Stratford, Connecticut. Depicted is the scene as log-book and documents for the first Seaguard are handed over. *(USCG)*

History has proved that Coast Guard aviators are great improvisors, and over the years have flown and evaluated many unique aircraft types, with the Flying Life Boat as the ultimate target to aim for. When the Sikorsky HH-3F Pelican helicopter is eventually retired, the Coast Guard will have no more amphibious aircraft. A successful Rescue Swimmer Programme is supposed to take up some of the slack in the missions, because there certainly are times when there is no substitute for reaching out and grabbing a survivor by the scruff of the neck — something that will be difficult to do from a low hover. Unless some extraordinary Service Life Extension Programme (SLEP) is developed, the Pelicans will reach the end of their economic service life by the end of the decade.

BUDGET BATTLE

The US Coast Guard is a complex organisation that operates scores of ships, aircraft, boats and shore stations. Coast Guard people perform many tasks — an aircraft may search for the survivors of a sunken ship, detect an oil slick, spot a vessel

An artist's concept of the Rockwell T-39 Sabre 715A fan-jet evaluated for the USCG during 1973/74, as a replacement for the Grumman HU-16E Albatross. It was one of many types including in the programme, and eventually the Fanjet Falcon was adopted, which became the HU-25 Guardian. *(USCG)*

smuggling drugs and rescue the survivors from a floundering raft — all during one sortie. The multi-mission approach allows a relatively small organisation respond to a wide variety of maritime needs. While the number of missions grew remarkably during the past decade, the number of people in the Coast Guard remains relatively constant. Reserves and volunteers in the Coast Guard Auxiliary routinely help with many tasks.

One must always remember that the Coast Guard is at all times an armed force of the United States, and is equal in such status to the US Army, US Navy, US Air Force and the US Marine Corps. This is dictated in Title 14, United States Code. In peacetime the USCG serves within the Department of Transportation. During conflict — or by Presidential decree — it reports to the US Navy Department. The service is decentralised administratively with headquarters in Washington, DC. Two area commands — on Governors Island, New York, and at Alameda, San Francisco — direct operations in the Atlantic and Pacific. Maintenance and Logistical Commands in those cities serve the area commanders and ten operational Coast Guard Districts.

Search and Rescue (SAR) is one of the Coast Guard's oldest missions; rescuing those in peril at sea has priority over all other peacetime missions. To minimise the loss of life, personal injury and property damage on the high seas, the Coast Guard maintains a nationwide system of boats, aircraft, cutters and rescue co-ordination centres on 24-hour alert, ready to respond to a vessel or aircraft in distress.

The Bell Model 222 helicopter was developed to meet the USCG requirement for a new SRR helicopter during 1978. It was demonstrated to the USCG Commandant in May 1977, and later to each USCG air station. On 13 March 1978 it was demonstrated at the USCG Aviation Training Centre, Mobile, Alabama. Depicted is the mock-up in full USCG livery.
(Bell Helicopter Textron)

As a result of Coast Guard SAR efforts between 1982 and 1986 more than 30,700 lives were saved, more than 707,000 persons were assisted, more than $3.8 billion of property saved and more than $11 billion of property protected. This cost the Federal government less than $2.4 billion. Defense Readiness ensures the Coast Guard can function as an effective armed force and its resources are used in both peacetime and wartime. During World War II 241,000 personnel including approximately 150 aircraft participated, with 3000 being involved in the United Nations Korean conflict, including six air detachments. In Vietnam 8000 US Coast Guard personnel manned vessels, cutters and patrol boats while Coast Guard aviators flew with the 37th Aerospace Rescue & Recovery Squadron (ARRS), performing rescue missions in hostile territory.

Annually the US Coast Guard faces a great strain as the Commandant testifies on Capitol Hill before the House of Representatives Sub-committee on Coast Guard on operational requirements covering expense funds in any Fiscal year budget. Dollar for dollar the Coast Guard returns as much or more to the US taxpayer annually than any other Federal Agency.

On 20 July 1978 Sikorsky released this photo of a mock-up of its entry into the USCG competition for a Short Range Recovery (SRR) helicopter. In June Sikorsky had offered to supply ninety twin-turbine S-76 helicopters to the service. On 26 March 1979 Sikorsky notified the USCG that it was withdrawing the S-76 Spirit from the competition. *(Sikorsky)*

The annual Coast Guard budget is normally divided under three main headings — Operating Expenses (OE); Research, Development, Test and Evaluation (RDT&E); Acquisition, Construction & Improvements (AC&I). Any one of these would include requirements for Coast Guard Aviation be it HH-65A maintenance and operation follow-on; air interdiction support; expansion of OPBAT (Operation Bahamas & Turks & Caicos); a Bahamas drug interdiction facility; land-based AEROSTAT (Mobile interdiction and surveillance radar system); HC-130H follow-on; acquisition of the new HH-60J Jayhawk medium range recovery (MRR) helicopter; HC-130H FLAR (forward looking airborne radar Phase IV, or building a hangar at Savannah air station in Georgia.

The Office of the Aeronautical Engineering Division includes a

During 1987, the US Navy and the USCG conducted flights of a leased Airship Industries SKS-500 airship. The 164ft long airship was equipped with surveillance radar, thermal image infra-red, an SAR winch, and an inflatable four-man power boat. The USCG evaluation took place at Elizabeth City. Depicted is a Goodyear airship *Mayflower* N38A in Coast Guard livery. *(USCG)*

systems manager for aviation computerised maintenance, this position the aircraft facility managers mentioned earlier, held by an experienced rated USCG pilot. Because of the horrendous lead time, Coast Guard Aviation is already looking ahead at potential candidates for its aviation fleet of the 21st century. Combining the range and speed of a fixed-wing aircraft with the hovering capability of a helicopter is the Bell-Boeing V-22 Osprey selected in 1983 to meet a requirement for a Joint Services Advanced Vertical Lift Aircraft (JVX), in which all US services, except the US Coast Guard had an interest. It is one of the most radical aircraft selected for production and service. It has generated great interest in USCG circles, and if appropriations for its full-scale production are forthcoming, then the Department of Defense (DoD) may include a number for the Coast Guard.

Such a concept would present the attractive possibility of replacing several types in the US Coast Guard Aviation inventory, so involving a single multi-purpose vehicle with which to accomplish the multi-role missions of the service.

On behalf of the US Coast Guard President Ronald Reagan requested $2.98 billion to operate the service in Fiscal Year 1989. This amount was required to help win the war against drugs, maintain military readiness and operate the world's most cost-effective search and rescue (SAR) team. Enforcement of Laws and Treaties began in 1790 when Alexander Hamilton formed a 'fleet of cutters' to suppress smuggling. Today, drug interdiction has made it one of the Coast Guard's most visible missions.

An artist's conception of the Bell-Boeing V-22 Osprey tilt-wing aircraft engaged in a Search & Rescue — SAR — operation. The great advance in aerospace technology demands the USCG keep abreast with any developments which may fit its operational envelope. *(Bell-Boeing)*

CHAPTER SIX
Coast Guard Aircraft Repair and Supply Center Elizabeth City

In the 1930s, aeronautical supply support for the US Coast Guard was achieved by the individual air stations requisitioning from the US Navy, or by purchasing direct from aircraft manufacturers. During World War II the air stations continued the pre-war supply support concept with larger reliance on the US Navy, since the Coast Guard was then operating mostly US Navy type aircraft such as the OS2U Kingfisher, PBM Mariner and PBY Catalina. With the end of World War II and the subsequent reduction from 150,000 to about 20,000 personnel, Coast Guard air stations were also reduced, leaving their shelves and storerooms heavily stocked with materials in excess of their needs. The idea of a central aviation supply centre for the Coast Guard was first conceived during 1945, and in the following year Elizabeth City in North Carolina, was

This excellent photo of a USCG General muster at Elizabeth City on 26 May 1950 depicts, in addition to the station personnel, no less than seven types in use by the service. They include one Curtiss R5C-1, one Martin P5M-, two Consolidated PBYs, one J4F- Widgeon, one JRF- Goose, two Boeing PB1Gs and a Sikorsky HO3S-1G helicopter. *(USCG)*

established and today is fully computerised to plan for the future needs of the service.

Although the idea for a central supply and overhaul unit for the Coast Guard was conceived early in 1945, it was not until the end of World War II that locations became available for an activity of this type. Among those available for consideration was the plant on Coast Guard property located in North Carolina. It had been constructed at Elizabeth City for the Consolidated Vultee Corporation by the US Navy, and designed and used as a modification centre for Lend-Lease aircraft being delivered to the Allies and had been closed down as soon as World War II ended. Elizabeth City had been the nerve centre of Coast Guard aeronautical activity for some time, so the redundant Consolidated plant was selected as the logical location.

In April 1946 a small complement of USCG officers and men were stationed as a sub-unit of the existing air station for the purpose of establishing the Coast Guard Aircraft Repair and Supply Base. After nine months of concentrated effort, the unit was ready to go into operation. On 3 January 1947 it was placed in commission as a headquarters unit with a complement of ten officers and sixty-three men. Although the supply function of the unit began immediately, it was not until October 1948 that the aircraft overhaul programme first produced results when a re-worked Grumman J4F Widgeon was delivered. It was then that the first civilian employees were used to supplement the military personnel.

During May 1964 the Aircraft Repair and Supply Base (ARSB)

Void of all paint and markings this Martin PBM-5G Mariner is seen waiting for a flight test after a complete re-work and overhaul at Elizabeth City during the 1950s. Today the huge organisation is computerised to deal efficiently with current USCG types. *(USCG)*

was renamed Aircraft Repair and Supply Center (ARSC). The fifty-five acre site includes two large hangars, a workshop and offices, a small hangar used for aircraft maintenance training and several frame buildings. There is a paint hangar and the units which occupy other buildings and hangars include the Repair Division's Industrial Machine Shop, the Support Center Public Works Division, the Warehouse-Management Information Services-Avionics Shop complex and Aviation Repair Division shops.

The overhaul, repair and modification programme at the ARSC, also referred to as the rework programme, has included the overhaul of over fifteen different types of aircraft and the modification of about six others. The rework is mainly accomplished by the Aviation Repair Division, with the support of the other divisions within the ARSC. In 1973 the modification portion of the programme was placed under the Aviation Engineering Division. From 1948 to 1953 the types of aircraft overhauled included the J4F Widgeon, JRF Goose, PBY Catalina, PBM Mariner, OY-1/2 Sentinel, HO3S, HTL, HO4S, and UF or HU-16E Albatross. By 1954 there had been established a continuing overhaul line for the UF Albatross and the HO4S helicopter, with occasional overhauls of PBM Mariners continuing. As the number of different types of aircraft in the Coast Guard inventory declined, the types being overhauled were naturally also reduced. In 1955 and early 1956 seven Grumman JRF Goose were overhauled with a further decrease in the number of UF and HO4S types. During this time, four JRB

Daily scene in the 1950s at Elizabeth City, North Carolina, showing three operational USCG types on the apron. They include Martin P5M-1 Marlin '1295', a Grumman JRF- Goose and a Boeing PB-1G Flying Fortress. Date of the photo was 19 February 1957. *(USCG)*

Expeditors, a single Douglas R4D, a PB-1G Flying Fortress and two Martin RM-1Z or VC-3A aircraft were overhauled.

By late 1958 the programme at the ARSC consisted almost entirely of the Grumman HU-16E Albatross and helicopter overhauls, the HU-16E being phased out during August 1979. Other types of aircraft have been overhauled sporadically, specifically the Fairchild C-123 Provider and an RM-1Z during the 1960/61 period. In 1964 the first Sikorsky HH-52A Seaguard was inducted and a transitions from Sikorsky HO4S or HH-19 to the HH-52A overhaul was made. In 1970 the first Sikorsky HH-3F Pelican was inducted, the overhaul programme building up to about fifteen HH-52A and twelve HH-3F overhauls a year. With the phasing out of the HH-52A the emphasis today is on overhauls to the HU-25 Guardian series and the HH-65A Dolphin helicopter. Over the years the drop in maintenance has been conducted on all aircraft including a Programmed Depot Maintenance (PDM) programme on the Convair HC-131 aircraft which commenced in March 1980.

In addition to aircraft overhaul, a number of electronic modifications and prototype installations have been made. In 1960 the provision of an APS-42 radar in the nose of the Fairchild C-123 was accomplished. This included a significant structural modification, as the blunt nose of the C-123 was replaced by the bulbous radome taken from Fairchild C-119 Flying Boxcar aircraft. Major modification of the single Lockheed EC-130E Hercules '1414' electronic equipment was also accomplished at Elizabeth City. In addition, other types such as the Douglas R5D (EC-54) Skymaster have been modified.

COMPONENTS

Aircraft components as well as aircraft themselves are overhauled. Parts are occasionally manufactured when this is more economical than purchasing outside, or when these parts are difficult to procure. An example of this is the fabrication of compound curvature windows and floorboards for helicopters. Components overhauled include generators, magnetos, and other electrical items, auxiliary servos, primary servos, main gearboxes, hand pump selector valves, main rotor heads, instruments of all types and avionic equipment. ARSC also has a complete engine rework and repair facility which ranges from the Allison T56 turboprop engine for the Lockheed Hercules to the General Electric engines for the Sikorsky HH-3F Pelican helicopter.

The Aviation Engineering Division comprises about forty highly qualified engineers and technicians who provide technical support for the entire Coast Guard Aviation community. A technical services branch reviews all incoming technical data for Coast Guard applicability and provides high priority technical assistance, particularly to air stations. Services include response

With two Douglas R5D Skymasters in the background the hull of a Grumman UF-2G Albatross is given a steam clean from personnel employed with the USCG at the Elizabeth City facility on 1 August 1958. *(USCG)*

to field questions on all aspects of aircraft maintenance, liaison with civilian contractors and US government agencies, technical research for maintenance and supply actions, investigating and documenting problems with maintenance and equipment and preparing interim time compliance technical orders (TCTOs), and other urgent information. To assist in these responsibilities, manufacturers' technical representatives for Coast Guard aircraft and engines currently on inventory are assigned to the staff.

A USCG projects branch is a descendant of the prototype branch, which was created in 1957. The branch prepared the prototype design for converting US Air Force SA-16 Albatross amphibians to Coast Guard UF- or HU-16. It made major changes in the Douglas R5D Skymaster, Sikorsky HO4S, HH-52A, HH-3F, the Lockheed C-130 and the Convair HC-131. It prepares and provides the TCTOs for aircraft modifications. The role was expanded to include preliminary engineering, prototype, kit design, and technical writing for major projects as well as some field team installations.

The support branch provides services concerned mainly with publications and documentation, including technical writing, drafting and photography. It maintains the Master Technical Library, helps to co-ordinate publication changes, generates supplements to major technical orders, provides pre-publication technical reviews for ARSC generated technical documents, and assists in the co-ordination of many projects and investigations. A quality assurance branch has the primary objective assuring that a quality product is put out by ARSC. Its inspectors cover all phases of the ARSC overhaul and repair facility, assigning highly experienced and motivated personnel to keep defects in the finished product to a minimum. Products procured from manufacturers are inspected to assure that manufacturing specifications are met, as are kits designed and assembled by the ARSC.

An Aviation Supply Division provides aviation material support for all Coast Guard air stations as well as Comptroller services to all Elizabeth City US Coast Guard units. As mentioned earlier, in the 1930s aeronautical supply support for the Coast Guard was achieved by the individual air stations requisitioning

Over fifty SA-16B amphibians of the US Air Force were transferred to the USCG in the late 1950s, all going to Elizabeth City for re-work and overhaul plus conversion. Depicted on 29 September 1959 is a hangar full of Albatross amphibs being converted including '17251'. A Douglas R5D-Skymaster is also noted in the hangar. *(USCG)*

either from the US Navy or by purchasing from aircraft manufacturers and aeronautical equipment vendors. This individual supply effort continued until 1946 when a central aviation supply activity for the US Coast Guard was conceived. The latest innovation is the Closed Loop Aeronautical Support System (CLASS) which monitors the location and condition of high-value avionics parts on the medium-range search (MRS) and short-range recovery (SRR) aircraft. An allowance of over a hundred military and civilian personnel provides a prompt, reliable and economic supply support to the expanding Coast Guard aircraft fleet and is accomplished with the most modern equipment and management techniques.

Over the years the engineers and technical experts at the ARSC have watched with great interest the evaluation of new types such as the Israeli Westwind, Cessna Citation (both of which were evaluated for six months in USCG livery), the Sabre 75A project, Fanjet Falcon, the Learjet and the British-manufactured Hawker Siddeley HS 125-600 and Hawker

The shell of a Grumman UF-2G Albatross amphibian '2130' and an engineless Fairchild C-123B Provider transport, are seen on the apron at Elizabeth City on 11 October 1961. *(USCG)*

Siddeley HS 748. Many types of helicopters were evaluated prior to the HH-65A Dolphin being adopted.

Looking back they can pride themselves in the many unique tasks completed, such as the conversion of the Grumman HU-16E Albatross to Coast Guard requirements, the earlier Sikorsky HO4S tugbird helicopter, which enabled USCG helicopters to tow ships, and modifications to Coast Guard helicopters for involvement in the huge NASA Gemini Missions for the retrieving of capsules and astronauts from the ocean after space missions.

During 1989 the Coast Guard acquired a handful of Sikorsky CH-3E rescue helicopters from the US Air Force and after rework and modification to USCG requirement, three were despatched to Traverse City air station, Michigan, while the remainder are held at Elizabeth City and use for support.

Apron view taken at the USCG Air Station Elizabeth City, North Carolina, on 29 February 1967, depicting no less than eight Lockheed HC-130 Hercules in the 1300 series, including '1339', '1340' and '1347', three Grumman HU-16E Albatross amphibians, and three Sikorsky HH-52A Seaguard helicopters. *(USCG)*

In contrast with the early years of hand-offs and hand-me-downs the Coast Guard is now equipped with a modern standardised fleet of very capable aircraft. Acquisitions which are presently under way or in the planning stages provide assurance that this will remain the case for the foreseeable future with the Aircraft Repair and Supply Center at Elizabeth City always ready to supply the necessary support.

AV TECH TRG CENTER

The residential USCG Aviation Technical Training Center opened at Elizabeth City on 4 August 1978. Initially its first graduates in 1979/80 entry level training was provided through four 'A' schools: Aviation electronics technicians, Aviation electricians' mates, Aviation structural mechanics and Aviation machinists' mates. With the addition in 1981 of an Aviation survivalman 'A' school, the training centre reached its full complement of entry level training for the five USCG aviation rates.

This training centre is also responsible for developing second generation training for the HH-65A Dolphin helicopter and from its inception has provided advanced aviation training through 'C' schools as needed by the Coast Guard. Currently this training covers the ADL-81 LORAN 'C' receiver, various radar units, aviation engineering administration covering logs and records,

Two Sikorsky HH-52A Seaguard helicopters, inlcuding '1392', are seen in the hangar at Elizabeth City during re-work and overhaul. Each aircraft is allocated a Works Order number — WO 1082 being allocated to HH-52A '1392. *(USCG)*

digital microprocessing, HH-3F Pelican automatic flight control and selected electrical maintenance, high reliability soldering, engine maintenance on the T58-5, T56 and T58-8B engines, to name just a handful of the variety of aviation aspects covered.

Located on the Pasquotank River in the rural Albermarle region of North Carolina, air station Elizabeth City has provided over fifty years of service to both the community and the Coast Guard. Following its commission on August 15 1940 the air station was under US Navy control during World War II and performed Search & Rescue as well as anti-submarine warfare duties. Since World War II the air station has become one of the largest in the service, and boasts the unique distinction of having the only Airport Control Tower manned by US Coast Guard personnel.

In Fiscal Year 1982 the air station flew 2,260 sorties, logging 5,340 accident-free flight hours. As a result the unit was credited with 35 lives saved, 265 persons assisted and more than $7,342,000 worth of property protected. In addition, law enforcement surveillance missions led to the seizure of more than a hundred tons of contraband.

Elizabeth City also serves as the Coast Guard's primary maintenance and aircrew standardisation unit for the ubiquitous Lockheed HC-130H Hercules. Area coverage ranges from the Eastern Seaboard of Canada to the Caribbean and the unit is today one of the largest and busiest in Coast Guard Aviation.

CHAPTER SEVEN
US Coast Guard Aviation Exhibit US National Museum of Naval Aviation

The huge US National Museum of Naval Aviation, located at the Naval Air Station, Pensacola, Florida, was officially established in December 1962 by authority of the Secretary of the US Navy. Initially it was contained in a small World War II temporary building. A major effort was initiated to collect the rapidly disappearing examples of historical aircraft, artifacts, and memorabilia of US Navy, US Marine Corps and US Coast Guard Aviation. The collection grew very quickly and the need for a more permanent and substantially larger home for the museum collection soon became apparent. In 1966, the Naval Aviation Museum Foundation was established to raise the necessary funds for a new museum building.

Phase I of the new museum was completed in April 1975, this being almost eight times the size of the original building. Through more hard work and dedication, Phase II of the museum was completed in 1978, bringing the total museum floor space to 107,000 square feet. All of this was accomplished with private capital raised by the Naval Aviation Foundation.

Shortly after the dedication of Phase II, the US Navy invited the US Coast Guard to join in the museum project. The Coast Guard promptly accepted the offer in as much as it has always shared a similar seagoing heritage and tradition. The Coast Guard also recognised that it had its own aviation story to tell — an exciting story of gallantry, quiet courage and sacrifice, in its multi-mission service on and over the oceans. The Coast Guard was invited to designate a trustee whose primary duties included fund raising. An advisory committee was also appointed to publicise museum objectives, attract artifacts and memorabilia and assist in the vital fund-raising efforts.

The advisory committee enlisted the support of a fledgling group of retired and active-duty Coast Guard aviators known as the Pterodactyls. This intrepid cadre of characters — led by some of Coast Guard aviation's best known, best loved and most talked about 'legends' — contributed and raised the

Scene on the USCG cutter *Northland* with Lt John A Pritchard Jr USCG pilot in heavy flight clothing watching while his Grumman J2F Duck is made ready for flight during November 1942. He was involved, with his crewman, in an attempt to rescue the crew of a Flying Fortress lost on a Greenland ice cap. Later the J2F Duck and crew were listed as missing in action. In 1943, Pritchard and Radioman B A Bottoms, received a posthumous award of the Distinguished Flying Cross. *(USCG)*

funds necessary ($100,000) to establish a Coast Guard exhibit within the museum. This exhibit, often referred to as a wing, was opened during impressive ceremonies, led by the then Commandant, Admiral James S Gracey USCG on 11 April 1983. The following day, Commander Elmer F Stone, Coast Guard Aviator No 1, was enshrined with appropriate and impressive ceremony, into the coveted Naval Aviation Hall of Honor. The second US Coast Guard aviator to be so honored was Captain B Macdiarmid, Coast Guard Aviator No 59. On 10 May 1989 Captain Frank A Ericson, a Coast Guard aviator since 1935, and a pioneer in helicopter development was included in the Hall of Honor.

PRESENT & FUTURE

The National Museum of Naval Aviation at Pensacola, Florida, has by now welcomed well over four million visitors from all nations through its doors. An impressive $7 million addition known as Phase III was officially opened by the US Navy Secretary, H Lawrence Garrett III, and the Secretary of Transpor-

Commander Frank A Erickson, USCG, Commanding Officer of the CGAS at Floyd Bennett Field, Brooklyn, New York, during World War II, seen congratulating Ensign Walter C Bolton USCG, for a job well done on his return from a rescue mission on 7 April 1944 in USCG HNS-1 helicopter BuNo 46445. *(USCG)*

tation, Samual B Skinner, on 12 October 1990. This has more than doubled the existing display area.

Today the US Coast Guard exhibit covers 3700 square feet and takes the visitor from the very first day of manned flight — as Coast Guard Surfmen from the Kill Devil Hills Life-saving Station helped Orville and Wilbur Wright launch their aircraft at nearby Kitty Hawk in 1903. These surfmen, stationed a little over a mile from Kitty Hawk, had provided logistic support for the Wright brothers and routinely assisted at the launching of their early experimental gliders. On that historic date — 17 December 1903 — Coast Guardsman John T Daniels, using a camera owned by the Wrights, took the only photo depicting the first flight of a heavier-than-air machine.

The forerunner of modern day Coast Guard served as midwives to the birth of manned flight. Photographs, murals and wall-mounted montages depict the beginning of US Coast Guard Aviation activities with scenes of Elmer Stone's early training at Pensacola in 1916, and his historic flight across the Atlantic Ocean as a pilot of the famous NC-4 in 1919.

Coast Guard aircraft of the 1920s are depicted in anti-smuggling activities, passing through to the 1930s when Search & Rescue (SAR) missions were initiated. During the 1940s as part of the US Navy, Coast Guard aircraft and crews operated on anti-submarine warfare (ASW) patrols and late in World War II trained aviators from Allied nations as helicopter pilots. The 1950s are well depicted, showing the extensive Coast Guard involvement in developing and improving the helicopter as both

Seen with the main rotors folded at the Pensacola museum is Sikorsky HO3S-1G '235'. It was donated by the Coast Guard during 1983, and represents one of the early post-war helicopters operated by the service up to 1957. *(US Navy)*

an ASW and SAR vehicle. The aircraft missions of the 1960s, 1970s and 1980s are well displayed in similar fashion with increasing emphasis on the ship-helo team and the multi-mission capabilities of the Coast Guard today.

An extensive collection of aircraft models and Coast Guard Cutter *Westwind* memorabilia are also included. There is a mini-theatre with a large screen TV showing Coast Guard Aviation videos. Currently five examples of Coast Guard aircraft types are located at the museum. These include Grumman G-44 J4F-1 Widgeon V212 which on the first day of August 1942 bombed and sunk the German submarine *U-166* in the Gulf of Mexico about 100 miles south of Houma, Louisiana. The Widgeon was piloted by Chief Aviation Pilot Henry Clark White, accompanied by his only crewman, Radioman First Class George Henderson Boggs Jr, and they were patrolling an assigned area near a buoy marking a sunken United Fruit ship. The aircraft and crew were based at Houma air station attached to Squadron 212 of the

This immaculate Grumman J4F-1 Widgeon 'V212' survived World War II during which its crew bombed and sunk German submarine 'U-166', was sold as war surplus by the USCG and after carrying three US civil registrations and as many owners, was received at the US Naval Aviation Museum in flyable condition during 1984. *(US Navy)*

Grumman HU-16E Albatross amphibian '7236' was received at the Pensacola museum during 1977 in flyable condition. It came from the USCG Air Station located at Traverse City, Michigan, and represents a great favourite operated by the service over many years. *(US Navy)*

Coast Guard. This was the only submarine sunk by Coast Guard Aviation during World War II. White was later awarded the Distinguished Flying Cross and Boggs received the Air Medal for his participation in the sinking of the 1120-ton Type IXC German submarine.

The famed Grumman J4F-1 Widgeon V212 survived World War II and was sold as war surplus to the US Department of the

During 1970 the US Naval Aviation Museum received this Grumman J2F-6 Duck BuNo 33581 from the US Naval Air Station at Norfolk, Virginia. It is now in Coast Guard World War II livery and represents one of the types operated during that conflict. *(US Navy)*

Seen parked outside the huge US Naval Aviation Museum complex at Pensacola, Florida, is this fully restored Sikorsky HH-52A Seaguard helicopter '1355'. It was carefully refurbished by personnel at the US Coast Guard Aviation Training Centre at Mobile, Alabama, and delivered to the museum in 1987. *(US Navy)*

US Coast Guard Aviation has its own pride of place in the huge US National Museum of Naval Aviation at Pensacola. This huge panel, one of many, places emphasis on SAR operations, depicting a Sikorsky HH-3F Pelican helicopter and USCG surfmen. In the foreground is the USCG designed rescue basket. *(USCG)*

Interior and over the years carried the US civil registrations N743, N2770A and N324BC. The aircraft was received by the US Naval Aviation Museum, prior to it being re-titled, in flyable condition during 1984 and restored in full Coast Guard livery.

A Grumman HU-16E Albatross 7236 was received by the museum in 1977 from the US Coast Guard Air Station located at Traverse City, Michigan, in flyable condition and represents one of the favourite amphibians operated by the Coast Guard over many years. The type can be found in Coast Guard livery in many museums and pleasure parks throughout the United States. During 1983 the Coast Guard donated a Sikorsky HO3S-1G helicopter 235 to the museum, this representing an early type used by the service. During 1970 the museum also received a Grumman J2F-6 Duck BuNo 33581 from the US Naval Air Station, Norfolk, Virginia. This is now in US Coast Guard markings, as is a Grumman JRF-3 Goose V190 which served at Brooklyn, New York, along with V191 during 1941.

During 1987, after many long hours of refurbishment by the men and women at the Coast Guard Aviation Training Center located at Mobile, Alabama, a Sikorsky HH-52A Seaguard helicopter 1355 was delivered to Pensacola for inclusion in the museum. The HH-52A now sits squarely in front of the entrance to the museum, proudly displaying its beautiful Coast Guard colours to all who enter or pass by. It will soon reside in the expanded area now being planned for the Coast Guard.

Recently the US Coast Guard Pterodactyls purchased a large-screen video for use in the Coast Guard wing of the museum. Plans are now being made to expand and improve the Coast Guard exhibit, and include an emphasis on ship-helo operations and drug air interdiction. Plans are also underway to increase the Coast Guard's representation in all National Museum of Naval Aviation events and activities including the semi-annual Foundation magazine.

To visit the museum is to savour eight decades of US Coast Guard Aviation history as seen from the vantage point of a special breed of aviator.

CHAPTER EIGHT
US Coast Guard Aviation Pterodactyls

The Ancient Order of the Pterodactyls has nothing to do with paleontology. It is a non-profit, fraternal and semi-professional organisation established to support and promote interest in Coast Guard Aviation and its history. It also encourages fellowship and *esprit de corps* among its members, who in the main are active duty and retired Coast Guard Aviators.

But why pterodactyls you may ask? It was decided at the founding of the order in the spring of 1977 that, because of the nature of the Coast Guard Aviation experience, the uniqueness of the pterodactyl made it an appropriate symbol of the organisation. This was originally worded 'The Pterodactyl, like Coast Guard Aviators, was unique, little recognised for its accomplishments, but highly respected by all who experienced its personal attention.' US Coast Guard Aviation is perhaps not as recognised for all its accomplishments as it could be, but it is highly respected by all who may be fortunate enough to see it in action and benefit from its services.

The Order itself, despite its short history, was deemed 'ancient' because of the Coast Guard's participation in the birth of aviation, when personnel from the nearby lifeboat station helped the Wright brothers make their memorable first flight a success. Roots that reach down into the very beginning of powered flight impart to the fraternal order the mystique of antiquity, without reflecting in any way on the age of members or aircraft.

The Ancient Order of the Pterodactyls was organised by a group of four retired US Coast Guard Aviators — Captain Andy Wall who was the originator, Captain George Thometz, Captain Gus Shrode and Commander Norm Horton — in Long Beach, California, in the spring of 1977. Membership is open to all persons who are serving or who have served honourably as pilots in Coast Guard aircraft, including those of other military services and foreign governments involved in exchange programmes with the Coast Guard. Associate memberships are tendered to individuals who have served in other capacities in Coast Guard aircraft, such as aircrewmen, flight surgeons, or personnel otherwise under official USCG flight orders. Membership by November 1990 stood at 719 members, comprising 607 regular members, 104 associates and eight honorary assignees. Regular membership includes pilots in the US Navy, US Air Force, Canadian Armed Forces and Royal Air Force.

The 1988 gathering was held at New Orleans, Louisiana; the 1989 gathering was held at Elizabeth City, North Carolina, while the 1990 meeting was held at the Experimental Aircraft Association (EAA) Museum, Oshkosh, Wisconsin. In order to ably celebrate the 75th Anniversary of US Coast Guard Aviation during 1991, the Pterodactyls will meet at the Naval Air Station Pensacola, Florida, the home of the US National Museum of Naval Aviation, the new name for the US Naval Aviation Museum with its tribute to both US Coast Guard veteran aviators and aircraft, now being enlarged with the space allocated being doubled.

CHAPTER NINE
Profile — Man and Machine

Elmer Fowler Stone

1887-1936
Commander, United States Coast Guard
Naval Aviator Number 38
Coast Guard Aviator Number 1

Elmer Fowler Stone was born at Livona, Livingston County, New York, on 22 January 1887. At the age of eight he moved with his parents, Frank and Elberta Fowler Stone, to Norfolk, Virginia, where Elmer graduated from Norfolk High School.

He was employed as a newspaper stereo typist prior to taking the entrance examination for cadets of the US Revenue Cutter Service, the forerunner of the US Coast Guard. Passing the examination as number one man, he was appointed as a Cadet on 30 April 1910. Three years later he graduated from the Academy and was commissioned a Third Lieutenant (Ensign) on 7 June 1913.

Subsequently he advanced in rank to Lieutenant (jg) on 27 June 1918, Lieutenant on 12 January 1923, Lieutenant Commander on 21 April 1924 and Commander on 1 May 1935.

His first assignment was to study the steam-propelled machinery on board the Revenue cutter *Onondaga*, based at Norfolk, Virginia, and a few months later he became a line officer in that vessel. After a temporary assignment on board the cadet training cutter *Itasca* from October 1914 to February 1915, he returned to the cutter *Onandaga*. Later he was commended by the Assistant Secretary of the Treasury for his skill and judgement in handling a lifeboat crew from that cutter in the rescue of seven crew members from the wrecked lumber-laden schooner *C.C. Wehrum* in gale force winds off False Cape, Virgina, on 4 June 1915. The shipwrecked seamen had been stranded without food for 48 hours.

Meanwhile, Elmer Stone had become an ardent advocate of aviation as a means of improving search and rescue (SAR) efforts. With a movement afoot in Congress to authorise a US Coast Guard Aviation branch, Stone put in a request for flight training early in 1916. As a result on 21 March 1916, he became

Rare photo depicting three of the crew of the world famous NC-4. *Left to right:* Lt (jg) W Hinton, US Navy, pilot; L/Cmdr A C Read, US Navy, aircraft commander; and Lt Elmer F Stone, USCG, pilot. *(USCG)*

one of the first two Coast Guard officers selected in the forming of the first Coast Guard aviation training group. Reporting for flight training at the Naval Air Station, Pensacola, Florida, by 1 April 1916, Stone qualified as a naval aviator on 22 March 1917 and on the following 10 April was designated Seaplane Aviator No 38 on the US Navy roster. It was not until 22 April 1920, however, that he was designated US Coast Guard Aviator No 1.

In July 1917, during World War I, Elmer Stone was assigned to the Aviation Unit on board the cruiser USS *Huntington* which was engaged in the convoy of troops to the battlefields in Europe. The cruiser carried a Curtiss R-6 observation seaplane plus Gallaudet and Martin hydroplanes. The R-6 became the first US-built aircraft to serve US forces overseas in World War I. However, despite earning the Victory Medal with Aviation Clasp for his tour of duty in World War I, Elmer became more and more convinced of the potential capabilities of aircraft for rescue work.

In October 1917, he was ordered to the Naval Air Station at Rockaway, Long Island, New York, a move that was to etch his name in history. It was while there as a First Lieutenant that

Elmer Stone became the only US Coast Guard member of US Navy Seaplane Division One selected to test the long-range capabilities of aircraft in an ocean crossing using three newly-designed NC model seaplanes, the NC-1, NC-3 and NC-4. The commissioning date was 3 May 1919. The NC-2 was scratched as its wings were used for the NC-1 which was damaged in a storm off Rockaway during a test flight. The unit began its journey from Rockaway beach at 10 am on 8 May 1919. The crew of the NC-4 comprised Lieutenant Commander Albert C Read, USN, Commanding; Lieutenant Elmer F Stone, USCG, Pilot; Lieutenant (jg) W Hinton, USN, Pilot; Ensign H C Rodd, USN, Radio Operator; Lieutenant J L Breese, USN, Reserve Pilot-Engineer; and Chief Machinist's Mate E C Rhodes, USN, Engineer.

The Curtiss NC-4 was the only one of the three aircraft to complete the historic first trans-Atlantic crossing, the other two flying boats having been forced down in fog and damaged at the gates of the Azores. When the NC-4 touched down in Lisbon Harbour, Portugal, on 27 May 1919, Lieutenant Stone, with the other crew members, that same day received the Portuguese government's decoration of the Knight of the Order of the Tower and Sword for valour, loyalty and merit upon successful completion of the trans-Atlantic flight. The NC-4 continued its flight to the United Kingdom where it arrived in Plymouth Harbour on 31 May. Elmer Stone received the British government's decoration of the British Air Force Cross (AFC). Later he received the US government's Navy Cross for distinguished service on 11 November 1920 and then the Congressional Medal for extraordinary achievement from President Hoover himself in the name of Congress on 23 May 1930. This medal was specially struck for the NC-4 crew. Stone also received a Silver Plaque presented by a French Scientific Society commemorating the NC-4 flight.

During July 1919, Elmer Stone reported for temporary duty (TDY) in the Aviation Section of the US Navy's Bureau of Construction and Repair in Washington, DC. That was followed by a brief spell as Executive Officer of the Coast Guard Cutter *Ossipee*. In November 1919 he was detailed to the outfitting of the cadet training cutter *Scally* at New London, Connecticut.

Returning to Washington, DC, in November 1920, he was detailed for six years with the US Navy Bureau of Aeronautics. During that period he served as test pilot and did considerable work on the development and installation of catapults and deck arresting devices for use on the new aircraft carriers *Lexington* and *Saratoga*, for which he was commended for valuable service by the Bureau Chief.

As history has so many times recorded, subsequent plans for establishing Coast Guard Aviation since World War I had been shelved. Elmer Stone still energetically supported attempts with a few other foresighted officers to initiate Coast Guard Aviation developments. The first attempt was the establishment of a Coast Guard air base at Morehead City, North Carolina, on 24 March 1920, which, though initially successful, failed after fifteen months owing to lack of funds. The second attempt during 1926 fully succeeded under the direction of Commander Carl C von Paulsen USCG.

In November 1926, after requesting to be returned to Coast Guard duty, Elmer Stone was ordered to sea duty as Executive Officer of the cutter *Modoc*, based at Wilmington, North Carolina. After two years on that cutter, he commanded first the destroyer *Monaghan* operating out of New London from September 1928 to June 1929, and then the destroyer *Cummings* until May 1931. Both these vessels were part of the US Coast Guard's Destroyer Force used in an all-out battle against rum runners and smugglers during the period 1924 to 1934.

During those years of sea duty Elmer Stone's interest in aviation never waned, even though many in service who had initially been interested dropped the matter. Fortunately USCG officers like Commander Norman B Hall who pioneered in aviation from the engineering standpoint, and Lieutenant Commander Carl C von Paulsen, an aviator, formed a unique triumvirate with Elmer Stone that directed the founding of the Aviation Branch of the US Coast Guard.

From May 1931 to March 1932, Commander Elmer F Stone served as a senior member of the trial board for new Coast Guard aircraft being built at the General Aviation Manufacturing Corporation at Baltimore, Maryland. These were Flying Life Boats known as FLBs, later designated PJ-1/2. For the next two years, Elmer Stone commanded the Coast Guard Aviation Unit located at Cape May, New Jersey. He participated in the search for survivors of the US Navy airship *Akron* which crashed off the coast of New Jersey in 1933. He was flying the Douglas RD-1 *Sirius* and when the US Navy blimp J-3, involved in the search for *Akron* survivors, also crashed, Elmer Stone landed nearby and recovered the body of the blimp's commanding officer.

Commander Elmer F Stone, USCG, was appointed a senior member of the board for the development of the General Aviation PJ-1/2 'Flying Life Boat' during 1931/32. Depicted is PJ-1 flying-boat at sea in the 1930s. The General Aviation Company was located at Baltimore, Maryland. *(USCG)*

The Douglas RD-1 *Sirius* amphibian was flown by Elmer Stone when he commanded the USCG Air Station at Cape May, New Jersey, in the search for survivors of the US Navy dirigible *Akron* in 1933. He picked up the body of the commanding officer of the blimp J-3 which also crashed during the *Akron* search operations. *(Peter M Bowers)*

During May 1934 Elmer Stone was ordered to Santa Monica, California, as inspector of Coast Guard aircraft being built by the Douglas Aircraft Company.

On 20 December 1934 Commander Stone established a world speed record for amphibian aircraft when he piloted a US Coast Guard Grumman JF-2 Duck '167' at a speed of 191.734 miles per hour over a three-kilometre test course at Buckroe Beach, Hampton Roads, Virginia. For that feat, Elmer Stone was commended by Secretary of the Treasury Morgenthau on 10

Grumman JF-2 Duck amphibian of the USCG, similar to the aircraft with which Elmer Stone established a speed record on 20 December 1934 at Buckroe Beach, Hampton Roads, Virginia, over a three kilometre test course at a speed of 191.734 miles per hour. *(William T Larkins)*

January 1935. He was also awarded a Certificate of Record by the National Aeronautics Association on 20 February 1935.

In May 1935 Commander Stone assumed command of the US Coast Guard Air Patrol Detachment located at the US Naval Air Station at San Diego, California. It was there that he met his untimely death a year later at the age of forty-nine. On 20 May 1936 while observing a new type of flying boat patrol aircraft that was undergoing US Navy trials, Commander Stone walked over and sat down on a concrete hangar abutment. He fell over, stricken with a heart attack that proved fatal almost instantly. Commander Elmer Fowler Stone was buried at Arlington National Cemetery. The first US Coast Guard Aviator was enshrined in the US Naval Aviation Museum's Hall of Honor at Pensacola, Florida, on 12 May 1983.

There was certainly more to Elmer Fowler Stone than his technical expertise. His ability as a leader is attested to by the way his subordinates remember him. 'He was the best skipper I ever knew', said Alvin Fisher, a First-class petty officer at the time, who later earned a Coast Guard commission. 'I was in the Coast Guard thirty-two years and I never met a commanding officer who cared so much for his men. He treated us like a father.'

Fisher had many memories of him and recalled vividly how Elmer Stone grilled Coast Guard student pilots assigned for flight training at Pensacola, Florida. He wanted to make sure they measured up to his standards and would make a good impression for Coast Guard Aviation. At the time Elmer Stone was commanding officer of the USCG air station at Cape May, New Jersey.

'All the Coast Guard officers who were headed for Pensacola had to go to Cape May first', Fisher said. 'When they got there, Elmer Stone would give them ten hours of preliminary flight training and then evaluate them. He was hard with them, but sincere, and he had their complete respect. He, in turn, respected the men under him and his fellow officers.'

On 26 January 1983, the USCG Group and air station new multipurpose building, housing Facilities Maintenance and Coast Guard Reserve at San Diego, was dedicated to the memory of Commander Elmer F. Stone USCG.

THE CURTISS NC-4 LONG-RANGING FLYING BOAT

Crossing the Atlantic has now become as routine as crossing the street. Every month thousands of aircraft of various types cross the fifty degrees west longitude line, on both east and west-bound flights. This was not so over seventy years ago and in May 1919, no man had made the crossing of the Atlantic by air, but many were preparing to try for the honour of being the first. It was a US Navy crew and aircraft, backed by the entire US Navy, who won the honour.

Apparently early in 1919, a young US Naval Aviator, Ensign Juan Terry Trippe, later the founder of Pan American World Airways, predicted that trans-oceanic flight was 'a perfectly safe and sane commercial proposition, not a gigantic gamble.' The US Navy NC-4 proved how right he was to be.

During September 1917 the chief of the US Navy's Construction Corps, Admiral David W Taylor, called in his key men, Commanders G C Westervelt, Holden C Richardson and Jerome C Hunsaker. These experts were in effect ordered to create what the combined efforts of the United Kingdom, France and Italy had been unable to achieve in three years of conflict — long range flying-boats capable of carrying adequate loads of bombs and depth charges as well as defensive armament sufficient to counteract the operations of enemy submarines. After this meeting, Glen H Curtiss was summond.

Within three days of his Washington, DC, meeting, Glenn Curtiss and his engineers had submitted general plans based on two different proposals. This design co-operation between the US Navy and Curtiss resulted in the famous NC flying-boats — Navy-Curtiss 1, 2, 3 and 4. The claim to fame of the NC series rests on the great accomplishments of a single example, the NC-4, the first aircraft to fly across the Atlantic Ocean.

One of the first problems not readily resolved was the aircraft's name. Westervelt applied the initials of his boss — DWT. Glenn Curtiss gave the designation TH-1 to the design, standing for Taylor-Hunsaker. It was Westervelt who changed it to 'Navy-Curtiss Number One' or simply NC-1. This was a logical compromise. Later under the revised US Naval designation system of 1923, all the NCs became P2N-1 for the Navy's second patrol design, at least on paper.

The group chose the Curtiss three-engined design with Commander Holden C Richardson called in to detail the design of the hull. Curtiss was to design the wings, empennage, and other items. The company finally received a contract on 27 November 1917. A month later it was decided to give Curtiss a contract to build four of the new flying-boats. Later, the Naval Aircraft Factory built six more.

Following the US Navy custom of numbering individual ships within a class, the designations NC-1 to NC-4 were assigned, being individual aircraft numbers for NC-class flying-boats. Construction of the NC-1 had commenced in Buffalo, New York, during December 1917. However, due to plant expansion design work on the experimental NCs received low priority from the new Curtiss management. It was not until the project was transferred to the new Experimental plant at Garden City, Long Island, that work did really progress.

A further problem arose as the existing Garden City factory was far too small for such an enormous construction contract. The US Navy built an enlarged shop, and also a new hangar capable of housing two assembled NCs at the Naval Air Station at Rockaway, some twenty miles from Garden City. The huge NCs were initially assembled at the Curtiss factory, then dismantled and trucked to Rockaway for reassembly and first flight.

There was also a manpower problem as all available skilled personnel were already employed in other factories located in the New York area, so a great deal of the actual construction of the NCs was contracted out to established boat builders and woodworking companies in the New England area. Soon after the completion of the NC-1 World War I came to an end, and at one time it did look as though the remaining NCs would be cancelled.

THE DREAM

At this time the longest recorded non-stop flight accomplished was about 1350 miles flown under ideal conditions and in the vicinity of a landing field. The suggested route across the Atlantic was over 1900 miles, over an area not well known for ideal flying weather — Newfoundland to Ireland. The idea was not new and as early as 1910 attempts were made to cross the Atlantic by air. First, there were balloons, non-rigid airships, which were successful only in provoking interest.

Prompted by foresight and good business sense, Lord Northcliffe, the British equivalent of William Randolph Hearst, announced a prize of £10,000 for the first successful transatlantic flight. Details were published in the London *Daily Mail* newspaper on 1 April 1913. French and Italian aviators were quick to enter while in the United States, Rodman Wanamaker, heir to the Philadelphia mercantile fortune, announced a contract with Glenn Curtiss to build a large flying boat. On 3 August 1914 Germany declared war on France and the next day Great Britain declared war on Germany. This was the beginning of World War I and the *Daily Mail's* prize was postponed.

After the end of hostilities many nations were preparing to compete for the *Daily Mail* prize for the transatlantic flight which was renewed. The US Navy decided to be the first to make the crossing, not for the money, but for the prestige of the US Navy. The project received official status on 4 February 1919. On 3 May 1919 Seaplane Division One was commissioned at Rockaway Naval Air Station, the brief plan being to use NC-1, NC-3, NC-4, and depart from Rockaway for Trepassey Bay, Newfoundland, some 950 miles distant that month. From there, there was a non-stop flight of 1381 miles to Horta in the Azores, a short 169-mile leg to Ponta Delgada then the 925 miles to Lisbon and finally a 500-mile leg to Plymouth.

The three Curtiss boats — NC-1 (BuNo A2291), NC-3 (BuNo A2293) and NC-4 (BuNo A2294) departed Rockaway for Trepassey on 8 May with departure for the Azores scheduled for 16 May. A string of twenty-one US Navy vessels were stationed along the route to provide radio communication and the necessary emergency assistance if needed. Due mainly to the problems of navigating in fog, only NC-4 completed the record 3925-mile flight, arriving at Plymouth on 31 May 1919.

CURTISS NC-4

With the hull constructed by the Herrescholl Company of Bristol, Rhode Island, the NC-4 duplicated the NC-3, and was launched on 30 April 1919. Under the command of Lieutenant Commander A C Read, USN it departed with the others for Trepassey, but was forced down with engine trouble and taxied sixty miles to the Naval Air Station at Chatham, Massachusetts. Following an engine change, it finally arrived at Trepassey on 10 May.

The Curtiss NC-4 US Navy flying-boat which made the first trans-Atlantic flight in May 1919, one of the pilots being Lt Elmer F Stone, USCG — Coast Guard Aviator No.1. *(USCG)*

Following an uneventful flight of nineteen hours twenty-three minutes, NC-4 reached Horta on 17 May. It then made the short flight to Ponta Delgada to await word from the missing NC-1 and NC-3. The NC-4 arrived in Lisbon on 27 May, reaching Plymouth on 31 May after an emergency stop at Mondego in Portugal, just north of Lisbon, resulting in an overnight stop at Ferrol in Spain, due to the delay.

Eight years before Charles E Lindbergh flew his epic solo flight aboard the *Spirit of St Louis*, the NC-4 proved the feasibility of trans-ocean flight, contrary to the beliefs of many sceptics at the time. The fifty-two hour trip, flown mostly through mist and drizzle, also demonstrated the potential of the US Navy's 2800lb flying boats.

Lieutenant Elmer Fowler Stone was the pioneer aviator of the

US Coast Guard. During World War I one of his many tasks in aviation had been as a test pilot, his performance earning him a place on the transatlantic list. As pilot of the NC-4 Elmer Stone knew everything about the flying boat and this knowledge contributed significantly to the success of the flight. It was Elmer Stone who assisted Chief Machinist's Mate Eugene 'Smokey' Rhoads USN and Lieutenant James Breeze USN who were responsible for repairing two of the NC-4's three engines when it was forced down near Chatham, Massachusetts en-route to Trepassey Bay.

The NC-1 and NC-3 were forced down by weather conditions and landed in stormy seas before reaching their destination, the Azores. The NC-1 sank while under tow by a ship that came to its aid. The crewmen of the NC-3 landed safely and drifted to the Azores, 205 miles away, and later taxied into the harbour under their own power.

NC-4 PRESERVED

After the transatlantic flight which took nineteen days and four flight legs to complete, the NC-4 was disassembled and shipped home to the United States, eventually to be turned over to the famous Smithsonian Institution. On return to the US it was reassembled and put on display in Central Park, New York. During November 1919, it was overhauled and flown on a recruiting tour for the US Navy.

Paul Garber, retired assistant director of the National Air & Space Museum, first saw the NC-4 at Rockaway Naval Air Station during the time of its preparation for the transatlantic flight. After the epic flight the saw the NC-4 twice in Washington, DC, once on the Anacostia River, and once on exhibit in the Washington Monument. He saw it one other time in Philadelphia.

Throughout his many years at the Smithsonian Institution, which he joined in 1920, Paul Garber retained his enthusiasm for aircraft. During World War II he was commissioned in the US Navy, and had already begun efforts to have the NC-4 brought to the National Air Museum for preservation and display. While in the US Navy he did try to keep track of the NC-4 which he believed was stored in the US Naval Gun Factory. However, he found the NC-4 in storage at the Naval Air Station, Norfolk, Virginia, in a building that was being emptied and the contents, including the NC-4, moved elsewhere. By more than good fortune, the commanding officer of Norfolk was Rear Admiral N L Bellinger USN who had been the pilot of the NC-1.

The admiral was instrumental in having the NC-4 moved into a new storage facility at Cheatham, Virginia. Here the flying boat remained until after World War II when Paul Garber arranged to have it moved to Washington, DC, and subsequently to the National Air & Space Museum's preservation and restoration facility in Silver Hill, Maryland. The hull of the flying boat had been earlier on display in the Smithsonian in 1920, but its size and the lack of facilities precluded display in its entirety.

During the early 1960s the restoration branch began work on the NC-4 for eventual display, but other priority projects eventually pushed the NC-4 restoration into the background until 1967 when the US Navy and the Smithsonian decided to refurbish it ready for display.

In 1969 it appeared on outdoor display in Washington on the 50th anniversary of the transatlantic flight by the NC-4. On 13 April 1975, the NC-4, on permanent loan from the National Air & Space Museum, was one of the twenty-three restored aircraft and thousands of ancillary exhibits in the newly dedicated US Naval Aviation Museum located at Pensacola, Florida. On 12 May 1983 Commander Elmer Fowler Stone USCG, pilot of the NC-4, was enshrined in the US Naval Aviation Museum Hall of Honor.

More than a tribute to man and machine . . .

AIRCRAFT PROFILES, ALPHABETICALLY LISTED

AÉROSPATIALE HH-65A DOLPHIN

On 13 November 1984 in a news release the US Department of Transportation reported that the US Coast Guard had announced plans for conditional acceptance of HH-65A Dolphin helicopters from the Aérospatiale Helicopter Corporation. The first of ninety-six helicopters ordered was to be delivered from the Aérospatiale assembly plant located at Grand Prairie, Texas.

It was during 1977 that an acquisition programme was launched to give the US Coast Guard a new short-range recovery (SRR) helicopter by late 1979 or early 1980. A Request for Technical Proposals (RFTP) was issued in September 1977, the proposal response from helicopter manufacturers being required by December 1977, with a Coast Guard decision on the new machine planned for mid-August 1978. The short programme cycle was due to the USCG, in consultation with the US Navy, in requiring all candidates to be able to receive a Federal Aviation Administration (FAA) derivative type certificate or a military equivalent in the acquisition time frame.

Helicopter manufactures who competed for the Coast Guard request requirement included Bell Helicopter Textron with a utility version of its Model 222, Sikorsky with a version of its

This HH-65A Dolphin '6544' was accepted by the USCG on 16 January 1987, its first operational Air Station being Savannah, Georgia. On 25 July 1988 it was transferred to Port Angeles, Washintgton, where this photo was taken. Cost of this useful helicopter is $3,067,000 and a total of ninety-six were procured. *(Ralph Peterson)*

S-76 Spirit, and Aérospatiale with a modified version of its SA 365. Aérospatiale conducted studies to see if the SA 365 would be a viable candidate for the programme. It was estimated it would have to install a pressure refuelling system and adapt the nose to accommodate a radar. The main change to the SA 365 was to be the replacement of the two Turboméca

Arriel engines to conform with the Buy America Act. A version of the helicopter had flown with US-manufactured Lycoming LTS 101 engines. The company proposed to enrol US vendors for the avionics, auxiliary power unit (APU) and the environmental controls. Aérospatiale emphasised the single-plot instrument flight rules ability, the hover stability and the fan-in-fin as great attributes to the SA 365.

In all its 75 years of aviation from 1916, the US Coast Guard has relied almost wholly for its aircraft upon those types already selected for service with either the US Navy, or in odd cases the US Air Force. The selection of the HH-65A helicopter in 1978 proved a major break with the tradition.

DESCRIPTION

Basically similar to the SA 365N already developed by Aérospatiale in France, the new Coast Guard version was identified by the manufacturers as the SA 366G and by the US Department of Defense with the designation HH-65A, and like the HU-25A Guardian was not used by any other of the US armed services. Its features included the Textron Lycoming engines, a Lucas full authority digital electronic control system (FADEC), Rockwell Collins avionics and a nose-mounted Northrop Sea Hawk forward-looking infra-red (FLIR) radar to aid rescue missions by night, in bad weather and in high sea states. Some sixty per cent of total cost was thus presented by US equipment, and final assemly took place at the Aérospatiale Helicopter Corporation located at Grand Prairie, Texas.

The first flight of a SA 366G was made in France on 23 July 1980 and the helicopter was later shipped to the United States to serve as one of the two trials aircraft used to development and certification of the HH-65A, this being obtained on 20 July 1982. The initial Coast Guard requirement for ninety Dolphin SRR helicopters was later increased by six, but introduction into service, intended for early 1982, was delayed by more than two years by a combination of problems.

Required specification for the Coast Guard's prime mission, Search & Rescue (SAR), stated that the short-range recovery candidate must take off, cruise outbound at 1000 feet in excess of 100 knots, travel 150 nautical miles from base, hover for thirty minutes and then pick up three 170lb survivors, return to base and have ten per cent of the available fuel, or twenty minutes of fuel remaining, whichever is greater, as reserve fuel at shutdown.

The first US Coast Guard air station to equip with the HH-65A Dolphin was New Orleans, Louisiana, during mid-1985. Primary task of this air station is to provide continual search and rescue (SAR) coverage in the Gulf of Mexico from Apalachicola, Florida, to the border of Lousiana and Texas. Other duties include environmental protection and increasing anti-drug trafficking patrols. The air station was selected to be the prime Dolphin unit because of the experience and competence of its personnel who are very professional.

DOLPHIN CREWS

All twenty-six officers are qualified pilots, and eighty-six out of 107 personnel are aircrew, performing flying duties in addition to their other functions. To maintain five pilots continually at readiness — three on unit and two at two hour standby — each of twenty-three duty pilots do eight 24-hour shifts per month. The aircrewmen also spend five days per month on 24-hour watch, and on an average USCG personnel work between sixty and eighty hours per week.

Aérospatiale HH-65A Dolphin conversion is thirty hours plus twenty hours on the night/dusk visual simulator, and is undertaken at the USCG Aviation Training Center (ATC) at Mobile, Alabama. Once at a front line unit such as New Orleans it takes 500 hours to reach first pilot status, and 700 hours before being checked out as an aircraft commander. Pilots with prior Coast Guard helicopter experience usually upgrade after 100 hours on the HH-65A Dolphin. Continuation training accounts for about 40 per cent of flight time, 25 per cent of which is done at night. Captains and first pilots must fly a minimum of forty-eight hours every six months. For co-pilots the minimum is twenty-four hours, but in practice active Dolphin pilots do far more. One of the largest source of new pilots for the Coast Guard to fly the HH-65A is from other US armed services and many are already rated military helicopter pilots on entry. NCO pilots from the US Army enter the Coast Guard with a commission under the scheme known in fact as Direct Commission Aviators.

The HH-65A Dolphin's sophisticated system is giving rise to a new breed of Coast Guard aviator. Much less hands-on flying is necessary, and some experienced USCG instructors are concerned that there is too much emphasis on button pushing during training at Mobile. To maintain piloting skills, the US Coast Guard had had to introduce a rule that at least 50 per cent of instrument approaches made with the Dolphin are flown manually.

New Orleans air station has some sixty enlisted aviation personnel who fly as HH-65A rear crew, and who are also responsible for unit maintenance. These are drawn from the five different Coast Guard aviation trades — engines, radios/electronics, wiring/lights, structural mechanics, and survival — and on each shift there will be one from each trade on duty.

Specialist trade training is undertaken at Elizabeth City, but HH-65A crew training is carried out at the unit. Most rear crew have basic first aid and there is alway a corpsman medic on duty, if not available one of the first aid trained mechanics is on station. Coming on line are the Aviation Survival Men —

survival equipment specialists, who are trained as swimmers and will be included in the Dolphin crew team.

SHIP DEPLOYMENT

New Orleans HH-65A helicopters make four annual ship deployments during which they assist in drug ediction operations. Dolphins of the US Coast Guard embark on both USCG cutters and US Navy ships; while the US Navy does not have the authority to board civilian vessels, it can provide an operating platform for USCG helicopters. Patrols last two to six weeks and are frequently carried out in the Bahamas and the passes of Yucatan. During trials with the cutter *Thetis* day landings with up to ten degrees of roll and night landings with up to seven degrees were successfully made.

Home-based HH-65A Dolphins are involved in the drugs war flying patrols out to forty miles off shore. The helicopters help to search and identify suspect vessels, but are not equipped for direct intervention. The unarmoured aircraft would be exceedingly vulnerable to any encounter with traffickers so they hold off out of small-arms range while directing USCG cutters in to make the boarding.

Drugs might be brought in on yachts, fast motor boats or innocent looking fishing boats. Most difficult to intercept are the 'stealth boats' — wooden vessels about 45ft long with a low freeboard, and apparently with radar-absorbing material around the engine compartment, making them almost impossible to detect.

SEARCH & RESCUE

During peacetime, one of the US Coast Guard's primary functions is SAR, and on average about 5700 lives are saved each year. At New Orleans in one year between September 1987 and October 1988 the contribution was forty-four lives saved with a further 151 people assisted, on 380 SAR missions performed. The air station requirement is to be able to put a minimum of two HH-65A Dolphins into the air at any one time. One is held at thirty minutes readiness, day or night, although in practice it is normally airborne in fifteen minutes. The second HH-65A is on two-hour standby. The rotors can be wound up in winds up to sixty knots and the new eleven-bladed epoxy fenestron tail unit can accept winds of up to thirty-five knots from any direction. A tannoy broadcast alerts the duty crew and gives task details, including any special requirements. The helicopter will be configured specifically for the mission, taking on additional fuel if required. A notable difference between the Sikorsky HH-3F Pelican and the HH-65A Dolphin is that the latter is short on range with a radius of action of only 150 miles, compared with twice that for the HH-3F. For longer range work, helicopters can refuel on the many oil rigs which are to be found up to sixty miles off shore. The HH-65A is also limited to a maximum of five survivors, so a second helicopter would be launched if greater capacity was required.

The SAR task is simplified if the vessel or aircraft in distress carries an Electronic Locator Transmitter (ELT). These transmit on VHF distress frequences 156.8 MHz (Channel 16), 121.5 MHz and 243.0 MHz, enabling the HH-65A to DF (direction finder) onto them. About ten per cent of call-outs are false alarms, and 75 per cent of these are caused by ELTs being activated in error. This will be reduced with the wider use of ELTs which give an encoded vessel or aircraft identity, so enabling verification. Sometimes false 'Maydays' are put out by drug traffickers to tie down USCG resources while a drug shipment is being made.

New Orleans is situated on the edge of the Mississippi Delta, an expanse of swamp that sprawls over an area 150 miles by 60 miles. It is dissected by a maze of channels in which substantial numbers of fishermen catch their living. Frequently the HH-65A helicopters are called out to locate an overdue fisherman, the area being so vast that without aerial assistance it might be up to a week before a small stranded boat was spotted. The task can require very extensive detective work by the HH-65A crew, particularly in deteriorating weather, and the Coast Guard crews have to use hand-drawn fisherman's maps giving the local Cajun names not found on official US charts. Some 25 per cent of tasks from New Orleans are conducted in the Delta area.

At a national level the Coast Guard responds to 11,000 oil spills annually, this being the figure for 1988, and operates two airmobile strike teams which undertake clean-up operations if the pollutor is unable to do so. Heavy tanker traffic to and from the oil refineries in Louisiana accounts for some 10-15 per cent of USCG operations from New Orleans, the task coming under Marine Environmental Protection. Although unable to clean up oil slicks, the Dolphin investigates reports, assist in identifying offenders, and airlift Coast Guard Marine Safety Officers or personnel from one of the USCG strikes teams to asses the situation.

ENGINEERING & OVERHAUL

The Aérospatiale HH-65A Dolphin is expected to serve with the US Coast Guard for the next 20-30 years, so it is imperative to adopt strict maintenance procedures. Designated the HH-65A prime unit, the personnel at New Orleans air station have played a major role in formulating these procedures. Odd serviceability problems have been experienced, such as the door falling off its roller on several occasions, but generally the teething problems were few. Inspection and checks are regularly carried out, maintenance philosophy being based on the Major Time

Compliance Order — time between component inspection. Compressor washes are made daily and deck plates are lifted every 400 hours to ensure corrosion is within limits.

Complete overhauls are carried out every thirty-six months at the Aircraft Repair and Supply Center located at Elizabeth City, North Carolina. Current average flying on the HH-65A is sixty hours per month. The fuel-efficient Textron Lycoming LTS-101 engine was selected to power the HH-65A instead of the original Arriel to satisfy US trade offset requirements. It apparently performed adequately on the prototype SA 366G, but since has received much publicity for poor serviceability with a figure of 61 per cent quoted for the Coast Guard fleet of Dolphins. The LTS-101 has to be re-bladed every 600 hours. The original air conditioning bled off seven per cent of engine power which the aircraft could ill afford. Consequently, a freon compressor system was used on the 53rd HH-65A, and earlier helicopters have been retrofitted.

Unfortunately the Dolphin is still underpowered, without a single-engine hover capability at most weights, and the manufacturer claims performance figures cannot be achieved within the engine's temperature limits. A redesign of the power turbine wheel has been considered, but sources within the Coast Guard indicate it cannot afford to retain the LTS-101. Funding has become available for two HH-65A Dolphins to be re-engined, and currently HH-65A '6556' was converted at Grand Prairie, Texas, during 1990 with T800 engines. The second HH-65A will be re-engined with Arriels. Flight test with the Allison/Garrett LHTEC T800 powerplants was expected to take place at Garrett's Phoenix, Arizona, facility. Early estimates indicate that the 6000-hour design lift could translate to an engine life of between 8000 and 9000 for the Coast Guard HH-65A application.

Seen with its original USCG serial '4109' is HH-65A accepted on 23 December 1986, and re-allocated serial '6539'. The Dolphin replaced the HH-52A as the USCG short-range helicopter. It has a range of 400 miles and is fitted with a sophisticated avionics package. *(USCG)*

The French-manufactured Arriel engine may have a tough time overcoming ingrained government prejudice. In testimony before a White House subcommittee early in 1990, former US Coast Guard Commandant Paul Yost said, 'I have made a rule of thumb as the commandant that I will never again buy a helicopter or an airplane that was not a DoD (Department of Defense) supported piece of equipment.'

Unfortunately the LTS-101 engine has had a troubled operational history since it began service in 1984. It has experienced numerous in-flight shut downs and has suffered from power turbine wheel cracking problems. The engine has operated at higher temperatures than initially designed. Result was that the engine, which was supposed to accumulate about 2400 hours between overhauls, has lasted less than 600 hours. Despite the poor reliability of the LTS-101 the reliability record of the HH-65A has been good. During 1989 the helicopter flew about 99 per cent of the hours expected, and in the first six months of 1990 it flew about 97 per cent of the hours programmed. The Coast Guard estimate for each HH-65A Dolphin in 1990 is 635 flight hours.

Textron Lycoming paid the US government a record $17.9 million and absorb a potential $60 million in maintenance and repair costs over the years up to 1996 to resolve government fraud charges over the performance of the LTS-101 engine fitted to the HH-65A.

THREAT

In time of conflict the US Coast Guard is transferred to the US Navy Department, and has primary responsibility for the Maritime Defense Zones (MDZs) including harbours and navigable waterways. The unarmed HH-65A Dolphins would be tasked with port security and anti-saboteur duties, perhaps deploying with a US Navy fleet task force, or to a cutter, for the protection of the MDZ. There has naturally been some discussion about arming the HH-65A for its war role. However, the Dolphin is considered to be too power-limited to carry a useful weapons load, although this could be changed if the re-engine programme is successful. As in previous conflicts, some US Coast Guard pilots would be seconded to the US Air Force on combat search and rescue duties.

For the foreseeable future however, Coast Guard helicopters including the HH-65A, will continue to perform a vital function in fighting the drugs war, acting as environmental guardians and standing by to defend the shores of the United States. There is no doubt that the HH-65A Dolphin should prove an excellent complement to the larger Sikorsky HH-60J Jayhawk now entering service in this truly multi-role task.

AÉROSPATIALE HH-65A DOLPHIN

Initially the US Coast Guard serials in the range 4104-4123 were allocated to the Aérospatiale HH-65A, with the USCG serials 4101 and 4102 allocated to the two trials helicopters which were not included in the Coast Guard contract production, but nevertheless appeared in US Coast Guard livery.

The following is a complete listing of the trials aircraft and the duplication of US Coast Guard serials which took place initially.

USCG Serial	c/n	Remarks
4101	6002	Transferred to the Israeli Defense Force.
4102	6003	Ground instruction airframe used for lightning protection tests.
4103	6006	Converted into Panther demonstration helicopter.
4104	6007	Transferred to the Israeli Defense Force.
4105	6025	Re-serialled 6543
4106	6032	6516
4107	6035	6553
4108	6038	6505
4109	6043	6538
4110	6044	6526
4111	6045	6539
4112	6048	6503
4113	6049	6501
4114	6050	6506
4115	6055	6502
4116	6056	6507
4117	6057	6504

Later the US Coast Guard re-allocated the serial range of the HH-65A helicopter to 6501-6596.

SERIAL INFORMATION

USCG No.	c/n	Commissioned		First unit assignment
6501	6049	19 Nov 1984	ex-4113	Miami, Florida
6502	6055	4 Jan 1985	ex-4115	Mobile, Alabama
6503	6048	17 Jan 1985	ex-4112	Mobile Alabama
6504	6057	15 Feb 1985	ex-4117	Mobile, Alabama
6505	6038	15 Apr 1985	ex-4108	Mobile, Alabama
6506	6050	10 Jun 1984	ex-4114	Elizabeth City, North Carolina
6507	6056	20 May 1985	ex-4116	Mobile, Alabama
6508	6095	29 Aug 1985		New Orleans, Louisiana
6509	6104	3 Jun 1985	temporary storage AMARC Davis-Monthan 47001	
6510	6105	24 Jul 1985		New Orleans, Louisiana
6511	6121	30 Aug 1984		New Orleans, Louisiana
6512	6123	12 Sep 1985		Boriquen, Puerto Rico
6513	6125	18 Sep 1985		Boriquen, Puerto Rico
6514	6127	30 Sep 1985		Boriquen, Puerto Rico
6515	6129	11 Oct 1985		Miami, Florida
6516	6032	31 Oct 1985	ex-4106	Miami, Florida
6517	6132	18 Oct 1985		Miami, Florida
6518	6135	3 Nov 1985		Miami, Florida

6519	6139	19 Dec 1985		New Orleans, Louisiana
6520	6120	19 Dec 1985		New Orleans, Louisiana
6521	6162	22 Jan 1986		Corpus Christi, Texas
6522	6164	17 Jan 1986		New Orleans, Louisiana
6523	6166	19 Feb 1986		Mobile, Alabama
6524	6167	3 Mar 1986		Mobile, Alabama
6525	6169	5 Mar 1986		New Orleans, Louisiana
6526	6171	21 Mar 1986		New Orleans, Louisiana
6527	6044	23 Apr 1986	ex-4110	Boriquen, Puerto Rico
6528	6172	11 Apr 1986		Mobile, Alabama
6529	6174	8 May 1986		Corpus Christi, Texas
6530	6175	22 May 1986		Corpus Christi, Texas
6531	6177	6 Jun 1986		San Diego, California
6532	6178	30 Jun 1986		San Diego, California
6533	6182	5 Aug 1986		Brooklyn, New York
6534	6183	17 Jul 1986		San Diego, California
6535	6185	29 Jul 1986		Brooklyn, New York
6536	6186	14 Aug 1986		Brooklyn, New York
6537	6188	19 Aug 1986		Brooklyn, New York
6538	6025	6 Oct 1986	ex-4105	Mobile, Alabama
6539	6043	23 Oct 1986	ex-4109	Cape May, New Jersey
6540	6189	8 Sep 1986		Brooklyn, New York
6541	6194	6 Oct 1986		Mobile, Alabama
6542	6195	24 Oct 1986		Los Angeles, California
6543	6045	6 Mar 1987	ex-4111	Savannah, Georgia
6544	6197	16 Jan 1987		Savannah, Georgia
6545	6198	15 Dec 1986		Miami, Florida
6546	6208	31 Oct 1986		Elizabeth City, North Carolina
6547	6221	9 Dec 1986		San Diego, California
6548	6222	15 Dec 1986		Cape May, New Jersey
6549	6223	23 Dec 1986		Humboldt Bay, California
6550	6238	16 Jan 1987		Mobile, Alabama
6551	6227	12 May 1987		Elizabeth City, North Carolina
6552	6243	16 Jan 1987		Cape May, New Jersey
6553	6035	11 Mar 1988	ex-4107	Elizabeth City, North Carolina
6554	623	8 May 1987		Mobile, Alabama
6555	6232	8 May 1987		New Orleans, Louisiana
6556	6235	4 Jun 1987		Miami, Florida
6557	6237	8 May 1987		Mobile, Alabama
6558	6238	11 Sep 1987		Mobile, Alabama
6559	6241	15 May 1987		Miami, Florida
6560	6243	15 May 1987		Mobile, Alabama
6561	6244	29 Sep 1987		Miami, Florida
6562	6247	26 Nov 1987		New Orleans, Louisiana
6563	6249	11 Sep 1987		Miami, Florida
6564	6250	29 Sep 1987		Mobile, Alabama
6565	6253	29 Sep 1987		New Orleans, Louisiana
6566	6257	30 Nov 1987		New Orleans, Louisiana
6567	6258	25 Nov 1987		Mobile, Alabama
6568	6259	23 Dec 1987		Elizabeth City, North Carolina
6569	6263	10 Dec 1987		Elizabeth City, North Carolina
6570	6265	10 Sep 1987		Mobile, Alabama
6571	6266	7 Dec 1987		Miami, Florida
6572	6267	11 Dec 1987		New Orleans, Louisiana
6573	6269	26 Feb 1988		Miami, Florida
6574	6270	17 Dec 1987		Miami, Florida
6575	6271	16 Feb 1988		Savannah, Georgia
6576	6272	7 Mar 1988		San Diego, Florida
6577	6274	25 Mar 1988		Collins Corporation

6578	6275	29 Jul 1988	Grumman Aero Corporation
6579	6277	28 Sep 1988	Corpus Christi, Texas
6580	6276	19 Jul 1988	San Diego, California
6581	6279	29 July 1988	Los Angeles, California
6582	6280	27 Sep 1988	Mobile, Alabama
6583	6281	31 Aug 1988	Savannah, Georgia
6584	6283	3 Nov 1988	Houston, Texas
6585	6284	25 Aug 1988	Cape May, New Jersey
6586	6285	31 Aug 1988	Brooklyn, New York
6587	6287	3 Nov 1988	Cape May, New Jersey
6588	6288	23 Dec 1988	Cape May, New Jersey
6589	6289	14 Apr 1988	Chicago, Illinois
6590	6291	21 Apr 1989	Cape May, New Jersey
6591	6292	20 Oct 1988	Elizabeth City, North Carolina
6592	6293	4 Nov 1988	Detroit, Michigan
6593	6295	23 Dec 1988	Detroit, Michigan
6594	6296	30 Jan 1989	Detroit, Michigan
6595	6297	27 Jan 1989	Chicago, Illinois
6596	6299	17 Mar 1989	Mobile, Alabama

US Coast Guard air stations known to have HH-65A Dolphins assigned include the following:
CGAS Astoria, Oregon; CGAS Barbers Point, Hawaii; CGAS Borinquen, Puerto Rico; CGAS Brooklyn, New York; CGAS Cape May, New Jersey; CGAS Corpus Christi, Texas; CGAS Detroit, Michigan; CGAS Elizabeth City, North Carolina; CGAS Chicago, Illinois; CGAS Humboldt Bay, California; CGAS Los Angeles, California; CGAS Miami, Florida; CGAS Mobile, Alabama; CGAS New Orleans, Louisiana; CGAS North Bend, Oregon; CGAS Port Angeles, Washington; CGAS San Diego, California.

TECHNICAL DATA

Manufacturer:	Aérospatiale
Other designations:	SA 365N
Span:	39 ft 2 in
Height:	13 ft 1 in
Empty weight:	5992 lb
Crew:	3
Top speed:	210 mph at sea level
Cruise:	160 mph at sea level
Engine:	Textron Lycoming LTS 101-750A-1
Take-off power:	2 × 680 shp
Rotor diameter:	39 ft 2 in
Designation:	HH-65A Dolphin
Type:	Short range recovery
Unit cost:	$3,067,000
Length:	44 ft 5 in
Rotor disc area:	1204 sq ft
Gross weight:	8928 lb
Passengers:	6
Range:	248 miles
Service ceiling:	7510 ft hover

BEECH JRB-4/5 EXPEDITOR

Developed from the civil Beech 18 or C-18 commercial light transport, it was two rather modest contracts in 1940 awarded to the Beech Aircraft Corporation of Wichita, Kansas, by the US Army Air Force, which were predecessors of more than 4000 examples built for many military air arms in the ensuing five years. The very first procurement was for eleven C-45s furnished as military staff transports with six seats.

Also ordered during 1940 were the first five of more than 1500 twin-engined Beech 18s for the US Navy and US Marine Corps which were allocated the designation JRB-1. Six more JRB-1s followed, and at the same time the US Navy ordered fifteen JRB-2 light transports. The JRB-1 was unusual in having a distinctive fairing fitted over the cockpit to improve all-round vision for the crew, and were intended primarily for photography. Initially known as the Beech Voyager, the official and more popular Expeditor followed as more orders were placed during World War II.

A total of twenty-three JRB-3s, similar to the US Army Air Force C-45B were ordered and fitted for photography, while the 328 JRB-4s were the US Navy equivalent of the standardised seven-seat UC-45F. During 1941 the USAAF adopted the Beech 18 as a navigation and bombardier trainer, the US Navy acquiring a quantity of similar aircraft for various training roles. These were equivalent to the USAAF AT-11, becoming the SNB-1 Kansan in 1942 with a dorsal turret and a modified nose fitted with a bomb aimer's position and used to train US Navy personnel allocated to patrol aircraft. Outwardly similar to the

Beech JRB-4 Expeditor BuNo 66469 seen at Watsonville, California, on 12 August 1956, shortly before final retirement of the type from the Coast Guard. A total of seven updated JRB-4/5 transports served between 1943 and 1956. A number served with the US Coast Guard Reserve.
(William T Larkins)

JRB-2, and equivalent to the USAAF AT-7, the SNB-2 Navigator was also ordered in 1942. There were variants such as the SNB-2C, SNB-2H and SNB-2P, indicating equipment and role changes.

Later models for the US Navy, built during World War II, included the updated JRB-4 and JRB-5 and between 1943 and 1958 the US Coast Guard operated seven of this ubiquitous Beech transport used mainly for administrative duties. The records indicate that the first batch of seven Beech SNB-1 JRB-4/5 transports were accepted early in 1943, with the final batch being delivered during 1947. One of the Coast Guard Beech transports was specially equipped to assist the Coast and Geodetic Survey department in harbour mapping and photography, just one of the many Coast Guard responsibilities today. Beech SNB-4 BuNo 90564 was based at Elizabeth City air station, North Carolina, and used on administrative duties and for proficiency flying.

It is confirmed that deliveries were commenced in April 1943, with the final one delivered in July 1947. Two were returned to the US Navy in 1948/49, the remaining five serving until 1956, but after that graciously retired. Beech JRB-4 BuNo 44605 served with the post-war Coast Guard Reserve. Two others identified from photographs include JRB-4s BuNo 66469 and 90580.

TECHNICAL DATA

Manufacturer:	Beech	Take-off power:	2 × 450 hp
Other designations:	C-45/AT-11	Designation:	SNB-1 JRB-4/5
Span:	47 ft 8 in	Type:	Trainer/transport
Height:	9 ft 4 in	Length:	34 ft 3 in
Empty weight:	6203 lb	Wing area:	349 sq ft
Crew:	2	Gross weight:	8000 lb
Top speed:	209 mph at sea level	Passengers:	6/8
Cruise:	117 mph	Range:	780 st miles
Sea-level climb:	1620 ft per min	Service ceiling:	21,500 ft
Engine:	Pratt & Whitney R985-AN3		

BELL HTL-1/4/5/7

A wide variety of early Bell helicopters were operated in small numbers by the US Coast Guard over the years 1947 to 1960, ranging from the HTL-1 to the HTL-7, and all used on a variety of tasks within the service. Two Bell HTL-1s were purchased in 1947 and used for survey fo the New York Harbor area under the direction of the Captain of the Port of New York. Checking for smuggling, harbour pollution, sabotage and other maritime derelictions, they usually had floats attached, although they were sometimes flown on skids. One HTL-1 in Coast Guard markings with US Navy BuNo 122461 had a wheel undercarriage. This was the last in the batch of ten used by the US Navy for service evaluation.

Photo evidence, so much responsible for some facets of US Coast Guard Aviation history, shows a Bell HTL-4 with US Navy BuNo 128623 and HU-1 unit marks and Coast Guard on the floats, appeared during July 1955. It operated from the USCG cutter *Storis* off Nome, Alaska, while the cutter was en-route to the DEW Line Operations in the Arctic. The helicopter transported vital mail and was used for ice reconnaissance flights, logistics support, general utility duties and search and rescue (SAR).

Three Bell HTL-5 helicopters were purchased by the Coast Guard, serving between 1952 and 1960. They were used on a variety of tasks, but their small size and short range limited their effectiveness. All three were commissioned from the Bell Helicopter Company located at Forth Worth, Texas, in February 1952.

Further photo evidence indicates that at least one Bell HTL-7 helicopter served on board US Coast Guard cutters, the photo indicating a hangar was available. Unfortunately USCG personnel obscure the USCG serial number.

HISTORY

A number of different versions of the well-known Bell 47 helicopter were procured for use by the US Navy between 1947 and 1958 commencing with the HTL-1. First flown on 8 December 1945, and granted the first-ever US type approval for a commercial helicopter in March 1946, the Bell Model 47 was still in production twenty years later for both civil and military customers.

Following the ten HTL-1 helicopters which were used for service evaluation and of which the Coast Guard obtained the last two in the batch, the US Navy ordered twelve HTL-2 helicopters in 1949. These two early models had a 178 hp Franklin 0-335-1 engine. Nine HTL-3s were delivered in 1950/1 with an uprated engine. The HTL-4 dispensed with the fabric-covered rear fuselage framework, although some were seen with and without covering. Deliveries to the US Navy during 1950/1 totalled forty-six, followed by thirty-six similar HTL-5s powered by a 0-335-5 engine.

The two final models for the US Navy was the HTL-6 delivered in 1955/6 and used for training helicopter pilots, while the HTL-7 was instrumented and also used for training.

Documentation discovered in USCG HQ archives indicates that two Bell HTL-7 helicopters, BuNo 145848 and 145853 were loaned by the US Navy to the Coast Guard from 1 August 1962, under Chief of Naval Operations letter OP 502D5/cr, Serial No 2698P50 to the Commandant US Coast Guard. During 1962 they were redesignated TH-13N and even later HH-13N and had a unit value of $163,969. A further letter dated 26 January 1967 from the Department of the Navy, Naval Air Systems Command, gave an extension of loan for these two helicopters until 1 October 1968.

On 6 December 1968, a requisition and invoice/shipping document was issued by the Commanding Officer, US Coast Guard air station, Route 1, Box 950, Warrenton, Oregon 97146 and signed on his behalf by Lieutenant (jg) W E Wade, USCG, Supply Officer, indicating that the helicopters BuNo 145848 and 145853 complete with floats and publications were being transferred to the Military Aircraft Storage Disposition Centre, Davis Monthan Air Force Base, Tucson, Arizona.

A USCG Bell HTL-4 BuNo 128623 lifts off from the deck of the USCG Cutter *Storis* off Nome, Alaska, during June 1955, taking mail to personnel working on the DEW Line operations in the Arctic. The type was also used on ice reconnaissance duties. *(USCG)*

TECHNICAL DATA

Manufacturer:	Bell	Tail-rotor diameter:	5 ft 8 in
Other designations:	H-13	Designation:	HTL-1/4/5/7
Contract No.	CG-19087	Type:	General purpose helicopter
Rotor diameter:	35 ft 2 in		
Height:	9 ft 4 in	Unit cost:	$49,290
Fuel:	29 galls	Length:	27 ft 4 in
Empty weight:	1435 lb	Blade area:	965 sq ft
Crew:	1	Oil:	2 galls
Top speed:	86 mph	Gross weight:	2350 lb
Cruise:	70 mph	Passengers:	2
Sea-level climb:	780 ft per min	Range:	212 miles
Engine:	Franklin 0-335-5	Service ceiling:	10,900 ft
Take-off power:	1 × 200 hp		

Miscellaneous information:
Unit value of HTL-7 (HH-13N) quoted as $163,696.

SERIAL INFORMATION

USCG No.	Commissioned	Decommissioned
Bell HTL-1	US Navy BuNo 122460	ex-USAAF YH-13 46-253 to USCG May 1947. Crashed December 1952.
Bell HTL-1	US Navy BuNo 122461	ex-USAAF YH-13 46-254 to USCG June 1947, disposed of January 1955
Bell HTL-4	US Navy BuNo 128623	
Bell HTL-5	USCG 1268, 1269, 1270	Commissioned February 1952
Bell HTL-7	US Navy BuNo 145858	
Bell HTL-7	US Navy BuNo 145853	

BELL HUL-1G

The four-seat Bell Model 47J helicopter with its lengthened cabin, was also procured by the US Navy and the US Coast Guard. Twenty-eight were delivered as HUL-1 in 1955/6 for general utility operations including service aboard icebreakers. The two Bell HTL-1 helicopters in use by the Coast Guard were replaced by two HUL-1Gs during 1959.

Two of the type were used for the first time by the Coast Guard in Alaska for ice reconnaissance work during the 1959 summer Bering Sea Patrol performed by the heavy duty icebreaker USCG *Northwind*. They were later assigned to the Coast Guard cutter *Storis* in the Bering Sea being used on search and rescue (SAR) and general reconnaissance duties.

Operable from either ship or land, the HUL-1G alighted on either skids or floats attached to each side of the fuselage. It is a four-place single-engine helicopter having a two-blade rotor and two-blade anti-torque rotor. The design also features a 400lb hydraulic hoist for rescue work and was fully instrumentated for night flying.

Seen parked at Kodiak USCG Air Station is Bell HUL-1G '1338', one of two operated by the service in Alaska. It was a variant of the civil Bell Model 47J. During July 1962, the HUL-1G was re-designated HH-13Q in a revision programme of type designations by the Department of Defense. *(USCG)*

The cabin was arranged to accommodate one pilot and three passengers, and the large transparent plastic bubble at the front of the cabin enabled the pilot to have a wide range of view. The engine, with a vertical drive to the main rotor, was located at the rear of the cabin.

USCG Bell HUL-1G 1337 fitted with floats operated from cutters and ice-breakers in the Gulf of Alaska area in the 1960s. Home base was Kodiak air station. Bell HUL1-G 1338 was fitted with skis and operated out of Kodiak air station during the same period. These two Bell helicopters served with the Coast Guard until 8 December 1967. Under a US Department of Defense directive dated 6 July 1962, all designations allocated to the US Navy were standardised with those of the US Air Force, so the HUL-1G became re-designated HH-13Q.

TECHNICAL DATA

Manufacturer:	Bell	Take-off power:	1 × 240 hp
Other designations:	Model 47J	Tail-rotor diameter:	5 ft 10 in
Rotor diameter:	37 ft 2 in	Designation:	HUL-1G
Height:	9 ft 4 in	Type:	Utility helicopter
Fuel:	48 galls	Length:	32 ft 5 in
Empty weight:	1618 lb	Gross weight:	2800 lb
Crew:	1	Passengers:	3
Top speed:	105	Ceiling:	15,000 ft
Cruise:	100 mph	Range:	200 miles
Sea-level climb:	1300 ft per min	Service ceiling:	17,000 ft
Engine:	Lycoming VO-435		

BOEING PB-1G FLYING FORTRESS

To supplement the rescue capabilities of US Coast Guard aviation, a total of eighteen ex-US Air Force Boeing B-17 Flying Fortress four-engined bombers were converted as 'flying life-boats' equipped with air droppable lifeboat fitted under the centre fuselage. These were converted and delivered during 1945. In addition to being operated on long-range search and rescue (SAR) missions, they were also used as patrol/observation aircraft with the International Ice Patrol off Newfoundland until 1957 when they were replaced by the Douglas R5D Skymaster. For twelve years one PB-1G of the USCG served as a special aerial photo mapping aircraft with the US Coast Guard and the US Geodetic Service. One source indicates that a further twenty-nine Boeing PB-1G aircraft were to be delivered to the USCG in 1946.

The history of the famous Boeing B-17 Flying Fortress is now a legend. Following a US Army Air Corps 'Circular Proposal' of April 1934 calling for a multi-engined long-range bomber it was on 26 September 1934 that Boeing engineers initiated work on the Model 299. This new project incorporated some features first tried out on the earlier Boeing Model 294, officially designated XBLR-1, later XB-15, a pioneering if unsuccessful design. The new Model 299 was smaller, more compact and had better defensive armament layout; it was also easier to build. The first B-17 prototype was rolled out on 17 July 1935, making its first flight eleven days later. This was the first of the Flying Fortresses, a term coined by a local newspaper reporter and later adopted by Boeing as their registered name. The rest of the success story has been adequately told elsewhere, but eventually 12,731 Flying Fortress bombers were built by Boeing, Lockheed and Douglas.

Although the US Army Air Force accepted 12,677 Boeing B-17 Flying Fortress bombers during World War II, the majority of these were either written off before 1945, or taken out of service and stored prior to going into the melting pot. When Strategic Air Command was formed in 1946 it was able to take on its inventory a few hundred B-17G aircraft, but these were either phased out or became transport versions so the type continued to serve in a variety of specialised roles for a number of years.

The first of these, adopted in fact during the closing stages of World War II, was as a useful air/sea rescue aircraft, with provision for a large lifeboat to be carried under the fuselage. Search radar replaced the chin turret and all other armament was removed. Plans were made for about 130 to be converted as B-17H, later designated SB-17G aircraft operating with Air Rescue Squadrons around the globe. These served in the European theatre until well into the 1950s.

By mid-1945 a requirement had arisen with both the US Navy and the US Coast Guard for a land-based long-range patrol aircraft. The US Navy required a radar-equipped anti-submarine and reconnaissance aircraft, while the US Coast Guard had the requirement for a long-range search and rescue (SAR) aircraft. On the last day of July 1945, the US Navy assigned the designation PB-1 to the Boeing B-17F and B-17G aircraft they intended to operate.

A number of factory-new Boeing B-17G bombers were held by the US Air Force in storage. These were ferried by the US Navy to the US Naval Aircraft Modification Unit — NAMU — located at Johnsville, Pennsylvania, for conversion. Primary changes for the aircraft intended for the US Navy was the installation of the large APS-20 search radar scanner in a fairing beneath the centre fuselage. Bomb-bays were sealed, extra fuel tanks installed, and these aircraft operated primarily in the early warning role, being designated PB-1W and given an overall blue finish. A total of thirty were eventually converted.

The US Coast Guard adopted the B-17 Flying Fortress during 1945 following the successful development by the US Air Force with the B-17H carrying a lifeboat externally. The eighteen ex-B-17G aircraft converted for the USCG were designated PB-1Gs. The last flight of a Coast Guard PB-1G took place on 14 October 1959. Photos show that at least one aircraft had its tail serial number prefix with a 'V', this being 77256 based at San Francisco air station during April 1947. Twelve PB-1G aircraft were still on the USCG inventory dated January 1950.

Boeing PB-1G 77254 was transferred to the USCG during September 1946, being unique in having only 52½ hours of flight time on the airframe. It was converted for photo survey

Boeing PB-1G Flying Fortress '7246' seen flying and equipped with a large flying lifeboat under the centre fuselage. A total of eighteen of these World War II bombers were converted and operated by the USCG between 1945 and 1957. In addition to SAR duties they were used by the International Ice Patrol. *(USCG)*

work, the bomb bay sealed and loaded with oxygen tanks. A massive hole was cut under the fuselage in order that a special US Coast & Geodetic survey $1,500,000 nine-lens aerial mapping camera could be installed. Only the World War II Norden bombsight remained to be used for pinpointing targets for the camera.

With the nine-lens camera, the PB-1G could photograph 313 square miles of terrain with one click of the shutter while operating at 21,780 feet. Most of the task was conducted at altitudes between 20,000 and 30,000 feet. For twelve years 77254 ranged from Puerto Rico to Alaska recording the face of the United States for the Hydrographic Office. During these years it flew almost an even 6000 hours and more than a million and a half miles. Her career was marred by two accidents, both occurring by coincidence at Hill Air Force Base, Ogden, Utah: one was a taxi-ing accident with a mobile tower being moved across the field which slightly damaged one wing, the other when she was struck by an unknown aircraft while parked. This damaged her nose.

As the years passed, the once familiar silhouette was greeted more and more with awe. She was the last of her type operated by the United States military services. Finally at 1.46 p.m. on Wednesday 14 October 1959 77254 made her final landing at Elizabeth City air station, North Carolina, bringing an end to yet another great era in Coast Guard aviation history. For twelve of her fourteen years of service with USCG aviation she had operated as a special photo mapping mount. She was finally broken up at Elizabeth City on 17 February 1960.

Records show that the odd Coast Guard PB-1G after becoming surplus to requirements was sold on the US civil aircraft market. During June 1959, 77250 was observed at Ontario, California, carrying the rough civil registration N4711C. Apparently it had previously been registered N8055E.

TECHNICAL DATA

Manufacturer:	Boeing	Take-off power:	4 × 1200 hp
Span:	103 ft 9 in	Designation:	PB-1G Flying Fortress
Height:	19 ft 1 in	Type:	Long range patrol and ASR
Empty weight:	36,135 lb		
Crew:	6	Length:	74 ft 4 in
Top speed:	310 mph at 25,000 ft	Wing area:	1420 sq ft
Cruise:	200 mph	Gross weight:	55,400 lb
Sea-level climb:	7 min to 5000 ft	Range:	2,500 miles
Engine:	Pratt & Whitney R-1820-97	Service ceiling:	35,000 ft

BOEING PB-1G FLYING FORTRESS

USCG No.	USAF Identity	c-n	USAF Serial
77245	B-17G-95-DL	32526	44-83885
77246	B-17G-110-VE	8721	44-85812
77247	B-17G-110-VE	8730	44-85821
77248	B-17G-110-VE	8731	44-85822
77249	B-17G-110-VE	8732	44-85823
77250	B-17G-110-VE	8733	44-85824
77251	B-17G-110-VE	8734	44-85825
77252	B-17G-110-VE	8735	44-85826
77253	B-17G-110-VE	8736	44-85827
77254	B-17G-110-VE	8737	44-85828
77255	B-17G-110-VE	8738	44-85829
77256	B-17G-110-VE	8739	44-85830
77257	B-17G-110-VE	8740	44-85831
82855	B-17G-110-VE	8746	44-85837
82856	B-17G-110-VE	8743	44-85834
82857	B-17G-110-VE	8747	44-85838

Notes: DL = Douglas, Long Beach. VE = Lockheed-Vega.

CASA C-212 300 AVIOCAR

During mid-1990 the US Coast Guard leased a single CASA-212 Series 300 utility transport, serial 0393 for suitability trials in support of helicopter operations. Painted in full USCG livery it was observed at the Miami Air Station located at Opa Locka airport, Florida.

The CASA C-212 Aviocar twin-turboprop light utility STOL transport was evolved by Construcciones Aeronauticas SA of Madrid, Spain, to fulfil a wide variety of military and civil roles; but primarily to replace the mixed fleet of aged and obsolete transport aircraft in service with the Spanish Air Force.

Initially the C-212 was proposed in four main versions — as a 16-seat paratroop transport, ambulance, military freighter, or 18-seat passenger transport, to be certificated to both military and civil standards. It had a STOL capability that enabled it to use unprepared landing strips of a minimum 1310 feet in length.

On 24 September 1968 CASA was awarded a contract by the Spanish Ministerio del Aire for the development and construction of two flying prototypes and an airframe for structural testing. During 1969 the detailed design was finalised, with over 500 hours of wind-tunnel tests carried out, and a full-size fuselage mock-up was built to evaluate such features as cabin vision, accessibility and location of the principal items of equipment.

CASA had initially a co-operative agreement with HFB of West

On the first day of July 1990, a single CASA 212-300 Aviocar '0393' was leased from CASA Aircraft USA. It was based at Opa Locka Air Station, Florida, on low-cost logistic support for operations in the Caribbean. A key feature is the high-volume cabin with rear loading cargo ramp. It has a STOL capability, is all-weather (IFR) equipped, and has a fully air-conditioned cabin. *(USCG)*

Germany whereby the Germany company contributed to the development and manufactured the wing centre-section for the Aviocar, including the engine nacelle, flaps and flap controls. The transport was also built by Nurtanio in Indonesia as well as CASA the parent company, in Spain.

The first flight of the CASA-212 took place on 26 March 1971. The original C-212-5 Series 100 Aviocar, of which 135 examples including the development aircraft, were built by CASA and twenty-nine under licence by Nurtanio (IPTN) in Indonesia, followed in 1979 by the improved 200 Series with more powerful TPE331-10 engines and increased maximum take-off weight. The standard model is the Series 300, which was certificated in December 1987. Military variants were designated C-212-M.

Certificated under FAR Part 25, the C-212 can be operated under FAR Part 121 and Part 135 conditions, and is well within the noise requirements of FAR Part 36. By January 1991, total sales of the Aviocar (all versions) had reached nearly 500 including 208 civil and 228 military aircraft, of which more than 420 had been delivered by CASA and IPTIN, with production continuing. This total includes thirty-seven Series 300 C-212-M for the air forces of Angloa (4); Panam (3); France (5); Boliva (1); Portugal (24); Colombia (3); Argentina (5); Leshoto (3). At least six civil Series 300s have been sold. The Aviocar is a simple yet effective transport, and the type's ruggedness and low costs has appealed to the Third World civil operators as well as military air arms.

The basic variants inlcude the C-212A military transport, the C-212-5 (later C-212 Series 100) civil type powered by 750 shp Garrett TPE5331-5-251C turboprops, the heavier C-212-10 (C-212 Series 200) with TPE331-10-501C turboprop engines, and the still heavier C-212 Series 300 with 900 shp TPE331-10R-513C engines driving Dowty Rotol four-bladed constant-speed fully-feathering reversible-pitch propellers. The current engines are equipped with an automatic power reserve (APR) system providing 925 shp in the event of one engine failing during take-off.

TECHNICAL DATA

Manufacturer:	CASA	Prop diameter:	9 ft 2 in
Other designations:	T-12	Designation:	C-212-300
Span:	66 ft 6.23 in	Type:	Utility transport
Height:	21 ft 7.72 in	Length:	52 ft 11¾ in
Fuel:	456 galls	Wing area:	430.56 sq ft
Empty weight:	9700 lb	Oil:	2 galls
Crew:	2	Gross weight:	17,637 lb
Top speed:	230 mph	Passengers:	24
Cruise:	220 mph at 12,000 ft	Stall speed:	90 mph
Sea-level climb:	1600 ft per min	Range:	519 nm with max payload
Engine:	Garrett Air Research TPF331-10R-5130	Service ceiling:	26,000 ft
		Prop blades:	4 Dowty Rotol
Take-off power:	2 × 900 shp		

CESSNA CITATION 500

The search for a replacement for its ageing fleet of Grumman HU-16E medium-range search (MRS) aircraft commenced by the Coast Guard in 1967. A major evaluation exercise involved more than thirty aircraft types whose performance was studied informally. These included foreign-built aircraft such as the Hawker Siddeley HS.748 and HS.125. By the early 1970s the new requirement for an MRS aircraft was more clearly established and turboprop, turbojet and turbofan types appeared to be prime candidates resulting in the operational evaluation fo a Grumman Gulfstream I, Israeli Aircraft Industries Westwind 1123, and a Cessna Model C-500 Citation.

Certain recommended candidate aircraft which included the Cessna Citation were leased by the US Coast Guard for a six month period to determine if the operational envelope and characteristics were suited to the Coast Guard MRS aircraft missions. Between 1 April and 30 September 1973 a detailed operational and technical evaluation was conducted with the aircraft home-based at the Coast Guard air station at Mobile, Alabama. Finished in full Coast Guard livery and with the tail serial CG 160 the Citation took part in deployments throughout the Coast Guard operating empire in the United States including Alaska. This was to introduce the aircraft to and evaluate it under different climes and conditions.

The MRS Task Organisation had Captain P A Hogue as its programme director, Commander B Harrington was the Project Manager, while Commander H J Harris Jr did the flight evaluation, assisted by Lieutenant Commander W C Donnell who did the engineering evaluation. Support activity at the home base of Mobile was organised by Captain A J Soreng, with operations handled by Commander J C Arney and engineering by Lieutenant G D Sickafoose.

It was on 7 October 1968 that Cessna first announced that it was developing a new eight-seat pressurised executive jet aircraft named Fanjet 500. After the first flight of the prototype on 15 September 1969 the aircraft name was changed to Citation. The second Citation flew on 23 January 1970 and by mid-February 1971, the two prototypes had accumulated almost 800 hours of flight in 600 sorties. The first production Citation 0001 N502CC made its first flight on 1 July 1971 and became the company demonstrator. Initial deliveries were scheduled for late 1971 certified to FAR Part 25 for transport category aircraft.

Neither the Cessna Citation or the IAI Westwind 1123 were selected to fill the MRS role, and after an abortive attempt to fill its requirement by purchasing Rockwell Sabre 75A aircraft through the US Navy the Coast Guard issued a RFP — requirement for purchase — to the US industry in January 1975. From a number of submissions, the choice was eventually broken down to a version of the Dassault-Breguet Mystere 20

Cessna Citation 500 '519' seen parked at Juneau Airport, Alaska, during the six month evaluation for a medium range search (MRS) aircraft during 1973. A replacement for the HU-16E Albatross was required. It was one of two types evaluated in full USCG livery. *(USCG)*

business jet, marketed in the United States by Falcon Jet Corporation and known as the Falcon 20.

The Coast Guard selected the Falcon 20G variant, designated HU-25A Guardian by the Department of Defense, although this was unique in that it was not selected by any of the other US military services. A contract for forty-one aircraft was confirmed on 5 January 1977. After evaluation the Cessna C-500 Citation was returned to the manufacturer.

TECHNICAL DATA

Manufacturer:	Cessna	Take-off power:	2 × 2200 lb thrust
Span:	43 ft 9 in	Designation:	Model C-500 Citation
Height:	14 ft 4 in	Type:	Medium-range surveillance
Fuel:	3403 lb		
Empty weight:	5456 lb	Length:	43 ft 6 in
Crew:	2	Wing area:	260 sq ft
Top speed:	330 mph	Gross weight:	11,500 lb
Cruise:	299 mph	Stall speed:	84 knots
Sea-level climb:	3100 ft per min	Range:	1200 nm
Engine:	Pratt & Whitney JT15D-1 turbofan	Service ceiling:	35,000 ft

CONSOLIDATED N4Y-1

Commissioned and procured by the Coast Guard during 1932 was a single Consolidated N4Y-1 single-engine land biplane. It was initially known as a Consolidated 21-A and inscribed as such on entering Coast Guard service. It was a commercial development of the US Army Air Corps PT-3 and the US Navy NY- trainer series, being produced for the Army Air Corps as the PT-11.

This single aircraft purchased under an Army Air Corps contract was initially powered by a 165 hp Wright J-6-5 engine, this soon being changed to a 220 hp Lycoming R-680 engine. It was only after the US Navy purchased three Lycoming-powered models of the Consolidated 21-C in 1934 that the Coast Guard model was re-designated N4Y-1.

Initially numbered CG-10, it later became 310 and even later V110. The N4Y-1 served until 1941 after being commissioned during August 1932. During 1939 it was based with the Coast Guard at Biloxi, Mississippi, and was also used as a training aircraft at Cape May, New Jersey.

TECHNICAL DATA

Manufacturer:	Consolidated	Rotor diameter:	9 ft
Other designations:	PT-11D -21C	Designation:	N4Y-1
Contract No.:	Army AC-4625	Type:	Trainer
Span:	31 ft 6 in	Unit cost:	$8000
Height:	9 ft 5 in	Length:	26 ft 11 in
Fuel:	903 galls	Wing area:	281 sq ft
Empty weight:	1862 lb	Oil:	9 galls
Crew:	2	Gross weight:	2544 lb
Top speed:	118 mph	Service ceiling:	13,700 ft hover
Engine:	Lycoming R-680-6	Two-blade prop blades:	Metal – fixed pitch
Take-off power:	1 × 200 hp		

A single Consolidated Model 21-A was operated by the US Coast Guard between 1932 and 1941, and used as a training aircraft at Cape May, New Jersey. It is depicted in the early days when it carried the serial 'CG-10', it later becoming '310' and later still 'V110'. It was purchased for the USCG by an Army Air Corps contract for $8000. *(Peter M Bowers)*

CONSOLIDATED PBY-5A/6A CATALINA

The first of many Consolidated Catalina flying-boats and amphibians to be used by the US Coast Guard was commissioned in March 1941 at San Francisco air station in California. This was the second production PBY-5 BuNo 2290 which became registered V189. During 1943 it went on TDY (temporary duty) to Alaska. By the end of World War II the Coast Guard was still operating 114 Catalinas which retained their US Navy BuNos and designations. It was 1954 before the last USCG PBY Catalina was finally phased out. The majority were operated on air/sea rescue (ASR) duties although one USCG patrol squadron — VP-6 — was based in Greenland throughout World War II on patrol duties. They were capable of carrying an airborne lifeboat under one wing, and could operate fitted with JATO bottles for an accelerated take-off.

During 1933 the US Navy ordered new flying-boat prototypes from both Douglas and Consolidated. The Douglas prototype XP3D-1 was not produced in quantity, while the Consolidated Model 28, designated XP3Y-1 became the most famous flying-boat and the most produced. It was built over ten years in six different locations, being operated by the US Navy, the US Air Force, the US Coast Guard, and many Allied air arms.

Designed by Isaac M Laddon, the Consolidated Model 28 had several distinctive features; these included a parasol mounted wing. Internal bracing made it possible to dispense with external struts. Except for two small members between the hull and the centre wing on each side, the wing was a true cantilever. The use of unique retractable stabilising floats which folded upwards to become the flush wing-tips in flight added to the aerodynamic characteristics. Two 825 hp Pratt & Whitney R-1830-54 engines were mounted on the wing leading edge. A tall cruciform tail unit was fitted. Up to 2000lb of bombs could be carried in addition to several .30 machine guns.

The US Navy ordered the XP3Y-1 on 28 October 1933 and the prototype, completed in the company's Buffalo factory, made its first flight on 15 March 1935. Early flight trials were successful and the US Navy ordered a initial batch of sixty on 29 June 1935, these being designated PBY-1 in the patrol-bomber category. The XPBY-1 was returned to Consolidated in October 1935, for modification up to production standards. This included the installation of R-1830-64 engines, plus a redesigned, less angular, fin and rudder. It flew again on 19 May 1936 and was delivered to a US Navy patrol squadron in October.

Production took place at the San Diego factory in California with contracts being placed for fifty PBY-2s on 25 July 1936, sixty-six PBY-3s on 27 November 1936 and twenty-three PBY-4s on 18 Decemer 1937. The PBY-3 had 900 hp R-1830-66 engines,

Ramp scene during World War II at San Francisco USCG Air Station, with the second production Consolidated PBY-5 BuNo 2290 carrying USCG serial 'V189'. It served during the years 1941/43 and saw service in Alaska. It was 1954 before the beloved Catalina was finally returned from USCG service. *(William T Larkins)*

and the PBY-4 1050 hp R-1830-72 engines. All these models were similar in appearance. On 20 December 1939 the US Navy ordered a further two hundred Catalinas which included 167 of the PBY-5 model, first deliveries of which were made on 18 September 1940. They had 1200 hp R-1830-92 engines and blister fairings over the waist gun position, as developed on the PBY-4. The fin and rudder was again redesigned.

In March 1941 the second production PBY-5 was commissioned by the US Coast Guard. It was still on the inventory of the San Francisco air station on 28 February 1943.

The first amphibious version of the Catalina design flew on 22 November 1939, it actually being the last PBY-4 built and redesignated XPBY-5A. The retractable tricycle undercarriage made very little difference to the performance, adding greatly to the versatility, so much so that the US Navy had the final thirty-three of the PBY-5s on order completed to the amphibian standard. A further 134 more PBY-5As were ordered on 25 November 1940, deliveries commencing at the end of 1941. When the United States entered World War II in December 1941 the Catalina was the principal patrol bomber flying-boat, and consequently played a prominent part in the Pacific theatre of operations.

Further contracts were placed with Consolidated during 1941/42 by the US Navy who procured 979 PBY-5s including one from a second production line at New Orleans, Louisiana, and 782 PBY-5As of which fifty-nine were built at New Orleans. Additional production began in Canada during 1941 at the Canadian Vickers plant located at Cartierville and the Boeing

Aircraft of Canada factory in Vancouver. The Canadian Vickers produced aircraft went to the US Army Air Force, others going to the Royal Canadian Air Force. Also during 1941, the Naval Aircraft Factory at Philadelphia received a contract to build 156 Catalinas. These were designated PBN-1 and named Nomad. A number of improvements were introduced including an increase in fuel capacity.

A product of the New Orleans factory was the PBY-6A, a version similar to the NAF PBN-1 but with search radar in a radome above the cockpit and amphibious undercarriage. These were ordered on 9 July 1943. An order for 900 was received but production ended after the delivery of forty-eight to the USSR, seventy-five to the US Air Force and one hundred and twelve to the US Navy. Production at New Orleans ceased in April 1945, bringing the total PBY-s built by the parent company to 2389. Grand total produced was no less than 3281 of which 1428 were amphibians.

OPERATIONS

In all theatres of operations during World War II the Catalina saw widespread service. Patrols were flown by night using aircraft finished in non-reflective black paint. These patrol squadrons were known as the Black Cats, first introduced by patrol squadron VP-12 from Guadalcanal in the Solomon Islands. In contrast, some of the world's worst weather was experienced by Catalina crews in the Aleutian Islands. The Attu Horny Bird symbol was adopted by the US Coast Guard air detachment based at Attu and equipped with PBY-5A Catalinas. Without doubt the most respected task, at least by rescued airmen and mariners, was that of 'Dumbo' air/sea rescue missions undertaken in nearly very ocean near the battlefront.

The US Coast Guard's long association with the International Ice Patrol and the Bering Sea Patrol made the service uniquely qualified for Arctic operations. Consequently during October 1941, Commander Edward H Smith USCG was appointed overall commander for Greenland defence, reporting to the Commander in Chief, Atlantic Fleet.

With the entry of the United States into the war, the air patrol requirement in the Greenland area was greatly expanded. From this requirement was born a special patrol squadron manned entirely by Coast Guard personnel and considered by many to be the most colourful of all the Coast Guard aviation units of World War II.

On 5 October 1943 Patrol Squadron Six (VP-6 CG) was officially established by the US Navy at Argentia, Newfoundland, relieving the US Navy's Bombing Squadron 126. The new squadron's home base was Narsarssuak, Greenland, code name Bluie West One (BW-1). Designated as a unit of the huge US Atlantic Fleet, it was under direct operational control of Commander Task Force Twenty-four (CTF-24) with administrative control vested in Commander Fleet Air Wing Nine (CFAW-9). All personnel matters, however, remained the responsibility of Coast Guard Headquarters.

Commander Donald B MacDiarmid, Coast Guard Aviator 59, a seasoned professional and flying boat expert, was selected to command VP-6. Thirty officers and 145 enlisted men were assigned; twenty-two of the officers were aviators and eight of the enlisted men were also pilots, most with considerable flying experience. All of the aircraft and ground crewmen had years of aviation service, every bit of which would be taxed to the limits during the more than two years of flying they would accomplish in the hostile environment of the North Atlantic.

The aircraft assignment called for the ten Consolidated PBY-5A Catalina patrol bombers, nine to be operational with one spare. However, because of delivery problems, flying operations commenced with only six aircraft. The PBY was a remarkable aircraft and quickly gained the respect and affection of both pilots and crewmen for its rugged dependability. It could carry 4000lb of bombs, two torpedoes or four 325lb depth charges. Cruising range at 105 knots was over 2000 miles. It did magnificent work and was employed for every conceivable mission. It could carry a crew of between seven and nine, and by 1945 most Coast Guard air stations were operating the Catalina.

The mission of VP-6 was five-fold; anti-submarine patrol, known as ASW — anti-submarine warfare; air support for North Atlantic convoys; search and rescue; surveying and reporting ice conditions; and delivering essential mail and medical supplies to military bases and civilian villages and outposts. German U-boats were operating almost at will in the North Atlantic and the number of convoy sinkings was staggering, giving VP-6's rescue duties high priority.

On 28 November 1943, not long after VP-6 had been commissioned, a US Army Air Force Beechcraft AT-7 twin-engined trainer was reported lost, and several Catalinas from Narsarssuak conducted a search over a wide area. Lieutenant A W Weuker finally located the wrecked aircraft on the edge of the Sukkertoppen Ice Cap on the first day of December. Six days later Weuker marked the spot with flag stakes; on 21 December photos were finally taken successfully to guide a rescue party. It was a USCG PBY-5A from VP-6 which directed the actual rescue party on 5 January 1944, on the first ten miles to the wreckage, dropped provisions to the rescuers, and two days later contacted the rescue group on the return trek and again dropped provisions.

As additional Catalina aircraft became available, the units area of operations broadened in scope and detachments were established at several locations. Two PBY-s and crews were based at Reykjavik, Iceland, furnishing air cover for both US Navy and Coast Guard vessels operating against the enemy and providing ASW services for North Atlantic convoys and search and rescue operations in conjunction with the Royal Air Force Coastal Command. While carrying out their missions, these units provided their own ground support.

An additional detachment of two Catalinas and crews was assigned to the Canadian Arctic in support of vessels entering the Hudson Bay area during the navigation season. Anti-submarine patrols were required in the Hudson Strait, Ungava Bay and Frobisher Bay area regions. Two more Catalinas were assigned on a rotational basis to the US Naval Air Facility at Argentia, Newfoundland, where all the major repairs and overhauls to the Catalinas were carried out. The widespread dispersal of aircraft and crews posed many administrative and logistical problems which made an already difficult situation even more unwieldy. But that was the hand VP-6 of the Coast Guard was dealt, and play it they did.

The operation was focused on Greenland, the largest island in the world, which lies almost entirely within the Arctic Circle. It is 1600 miles north to south and nearly 800 miles wide. Eighty-five per cent of the island is covered with a great ice cap of unbelievable thickness. It was not uncommon for VP-6 aircraft to fly thousands of miles over the ice cap under the most trying weather conditions in a single search. Strong winds of 120-150 knots were a constant threat. Flying in those weather conditions, far from bases and with few navigation aids, required a high degree of pilot skill and courage. Only well trained, savvy pilots and crews could have survived.

Hundreds of rescues were carried out by VP-6 during its twenty-seven months of operations, frequently during high winds and near zero visibility. During a three-month period early in 1944, Lieutenant Carl H Allen, USCG, flew more than a hundred hours each month over difficult Arctic terrain to and from convoy support duty and ice patrol. One flight took him over the magnetic North Pole.

SERIAL INFORMATION

The Consolidated PBY-5A Catalina saw wartime service with the US Coast Guard between 1942-45 and by the end of 1944 no less than 114 were in service. These were gradually phased out during 1954. Six improved PBY-6A models were acquired in 1945, but were mustered out in 1954.

At San Francisco air station in California, the single PBY-5 V189 ex-US Navy BuNo 2290 was still on inventory on 28 February 1943, along with a single PBY-5A. On 31 October 1944 the air stations Catalina complement had risen to three PBY-5A amphibians.

The status of USCG aircraft dated 31 October 1944 listed the location of PBY-5A Catalinas as follows:

Elizabeth City, North Carolina	5

Port Angles, Washington	2
St Petersburg, Florida	3
Salem, Massachusetts	1
San Diego, California	4
San Francisco, California	3

By 1 May 1945 the stock of USCG PBY-5A Catalinas had been reduced to fifty-six, and by January 1950, to thirty-two. One of the six PBY-6A amphibians was BuNo 64096 designated PBY-6AG and seen at San Francisco in July 1952. An unusual PBY-5A also seen at San Francisco on 3 April 1947 had the BuNo V48281, while a World War II PBY-5A was BuNo 7293 with 'FRISCO-D' on the front of the fuselage.

Five PBY-5A Catalinas of VP-6 Squadron have been identified, these being BuNo 2464, 2470, 46510, 46572 and 46575. Two more USCG PBY-5As — unit unidentified — include BuNo 05021 and 34020.

On 15 May 1944 Commander William I Swanston, USCG, relieved Commander MacDiarmid as commanding officer of VP-6, and Lieutenant Commander G. Russell 'Bobo' Evans, USCG, became executive officer. By then the squadron had twelve PBY-5A Catalina aircraft, with two in Iceland, two assigned to the Canadian Arctic, three at Argentia and five at BW-1 Narsarssuak. Throughout the summer of 1944 the squadron was extremely busy. An expanded part of the squadron's operations involved ASW operations in the Baffin Bay, Davis Strait and Labrador Sea areas to protect US ships transporting cryolite, urgently needed in the protection of aluminium for the US aircraft production programme.

By early 1944, rapid expansion of US Coast Guard aviation had produced a shortage of seasoned pilots and crews. To provide some relief for VP-6, a pre-training schedule syllabus was introduced at Elizabeth City air station, North Carolina. Coast Guard HQ decreed that a one-year tour of duty in the Arctic regions was sufficient, and so ordered that pilots and crews not requesting an extension of their tour be relieved as soon as possible.

To comply with a US Navy directive dated the first day of October 1944, all patrol squadrons (VP) and multi-engine bombing squadrons (VB) were renamed and redesignated patrol bombing squadrons. Thus VP-6 (CG) became VPB-6 (CG).

The surrender of Germany in May 1945 brought U-boat activity to a standstill. However, VPB-6's operations in search and rescue (SAR), ice patrol, logistic support of military bases, LORAN stations and civilian facilities continued unabated. On 30 May 1945 Commander Loren H Seeger, USCG, relieved Commander Swanston as commanding officer of VPB-6. On 12 July administrative control of the squadron was transferred from Commander Fleet Air Wing Nine to Commandant US Coast Guard and it was redesignated a non-combat squadron. However operational control was retained by Commander Task

Seen at San Francisco USCG Air Station during July 1952 is a visiting Consolidated PBY-6AG BuNo 64096 in full Coast Guard livery and having a broad yellow RESCUE band on the rear fuselage. Over 120 of the type served between 1941 and 1954, and it equipped a USCG squadron — VP-6 — in Greenland during World War II. *(Douglas D Olson via W T Larkins)*

Force Twenty-four. In August 1945 VBP-6 received a directive to transfer its base headquarters from BW-1 Narsarssuak to the US Navy facility at Argentia, Newfoundland, where it was disestablished as a US Navy squadron in January 1946.

More detailed accounts of this unique squadron can be found in *A History of US Coast Guard Aviation*. The dedication of all members of the squadron have added a heroic chapter to the story of US Coast Guard Aviation.

TECHNICAL DATA

Manufacturer:	Consolidated	Designation:	PBY-5 Catalina
Other designations:	Model 28	Type:	Patrol flying-boat
Span:	104 ft	Unit cost:	$182,668
Height:	18 ft 11 in	Length:	63 ft 10 in
Fuel:	1750 galls	Wing area:	1400 sq ft
Empty weight:	15,384 lb	Oil:	116 galls
Crew:	10	Gross weight:	24,332 lb
Top speed:	202 mph at 6000 ft	Stall speed:	69 mph
Cruise:	125 mph	Range:	4000 miles at 6000 ft
Sea-level climb:	1130 ft per min	Service ceiling:	23,400 ft
Engine:	Wright R1830-82	Three-blade prop blades:	Curtiss-controlled pitch
Take-off power:	2 × 1200 hp		
Prop diameter:	11 ft		

TECHNICAL DATA

Manufacturer:	Consolidated	Take-off power:	2 × 1200 hp
Span:	104 ft	Designation:	PBY-5A/6A Catalina
Height:	20 ft 2 in	Type:	Patrol flying-boat
Empty weight:	20,910 lb	Length:	63 ft 10 in
Top speed:	175 mph at 7000 ft	Wing area:	1400 sq ft
Cruise:	113 mph	Gross weight:	35,420 lb
Sea-level climb:	620 ft per min	Range:	2350 miles
Engine:	Wright R1830-92	Service ceiling:	13,000 ft

CONSOLIDATED P4Y-1 LIBERATOR

The first batch of Consolidated PB4Y-1 four-engine, twin-tail, Liberators were delivered to the US Coast Guard early in 1944, the remainder following a year later in 1945. The total acquired was five which were operated by the USCG in the rapidly expanding search and rescue (SAR) effort during and immediately after World War II. Their long range and reliability made them outstanding platforms for extended ocean searches far from shore. They were all disposed of in 1951.

Design of the Consolidated B-24 Liberator bomber commenced early in 1939, when the US Army Air Corps outlined to Consolidated a requirement for a heavy bomber of even better performance than the Boeing B-17 then in production. A better range was required in the region of 3000 miles, together with an increase of speeds in excess if possible of 300 mph and a ceiling of 35,000 feet.

It is not generally realised that more B-24 Liberators were built for the US Army Air Forces, the US Navy and the Allied air forces during World War II, than any other single type of US aircraft. Overall total for the USAAF alone amounted to 18,325. It did not enter service until 1941, was withdrawn from service in 1945, and had a most remarkable record. It was built by Consolidated at San Diego, California, and Fort Worth, Texas, and by Douglas at Tulsa, Oklahoma. At the end of 1942 a fourth production line was provided by the Ford Motor Company in a new factory at Willow Run, Michigan, while North American at Dallas, Texas, began deliveries from their production line, the fifth, during 1943. Like the Boeing B-17 Flying Fortress, the global history of the B-24 bomber has already been catalogued for posterity.

During World War II the US Navy experienced limitations with flying-boat patrol bombers, as well as advantages. It was in 1942 that the US Navy first requested a force of B-24 Liberator land-based bombers to fly long-range over water patrols against shipping and submarines. An agreement was reached on 7 July 1942, allowing the US Navy a share of B-24 production.

It was August 1942 before versions of the B-24D, designated PB4Y-1, began to reach US Navy formations. The first operational unit was based in Iceland, this scoring its first success against a U-boat on 5 November 1942. After a slow build up with deliveries, the USAAF in August 1943 agreed to transfer over its anti-submarine B-24 squadron to the US Navy, and accordingly disbanded the Anti-Submarine Command at the end of that month. The ASV-equipped B-24s were transferred to the US Navy as PB4Y-1s in exchange for an equal number of unmodified Liberators already on the production line for the US Navy. A total of 997 PB4Y-1s were built plus some transport versions of the Liberator, designated RY-1 and RY-2.

A number of US Navy Liberators were modified for reconnais-

Five four-engined World War II Liberator bombers, designated initially P4Y-1 and used as long-range patrol bombers were operated by the USCG. They served the USCG from 1944 to 1951. US Navy records reveal that nine were operated by the Coast Guard, these being re-designated P4Y-1P just prior to retirement and disposal. *(AP Photo Library)*

sance duties, being designated PB4Y-1Ps and serving until the 1950s. In 1951 they were redesignated P4Y-1Ps shortly before being withdrawn from service. The records indicate that the nine US Coast Guard PB4Y-1s were also redesignated just prior to disposal, also becoming P4Y-1s. Unfortunately no details of their service with the Coast Guard are available.

TECHNICAL DATA

Manufacturer:	Consolidated	Take-off power:	4 × 1200 hp
Other designations:	B-24	Designation:	PB4Y-1/P4Y/1 Liberator
Span:	110 ft	Type:	Patrol
Height:	17 ft 11 in	Length:	67 ft 3 in
Empty weight:	36,950 lb	Wing area:	1048 sq ft
Crew:	6	Gross weight:	60,000 lb
Top speed:	279 mph at 26,500 ft	Range:	2960 st miles
Cruise:	148 mph at 1500 ft	Service ceiling:	31,800 ft
Sea-level climb:	830 ft per min		
Engine:	Pratt & Whitney R-1830-43 or 65		

CONVAIR PB2Y-3/5 CORONADO

Two huge Consolidated PB2Y Coronado four-engine twin-tail flying-boats were acquired by the US Coast Guard early in 1944, followed by the remainder of five in 1945. These flying boats were used by the USCG on anti-submarine warfare (ASW) patrols and for long-range search and rescue (SAR) missions. They were a mixture of the PB2Y-3 and the PB2Y-5 models fitted with low-altitude R-1830 engines and with an increased fuel capacity. The odd PB2Y-5H was also used by the Coast Guard, this having cabin accommodation for twenty-five stretchers. They operated in small numbers from the air station located at San Francisco, California, and the type was withdrawn from USCG service during 1946.

The Consolidated Aircraft Corporation of San Diego, California, were very involved in flying-boat design prior to World War II, in conjunction with the US Navy. In less than three months of the first flight of the Consolidated XP3Y-1 prototype of the versatile Catalina, the US Navy began making plans for the development of much larger flying boats with even better operational performance. In addition to Consolidated, prototypes were also ordered from Sikorsky, which appeared respectively in June 1935 and July 1936, for a large four-engine flying-boat in the patrol bomber category.

Consolidated produced the Model 29, making use of retractable wingtip floats similar to those fitted to the PBY Catalina. Other than that it was a wholly new design with a high-mounted wing and a capacious hull with accommodation for a crew of ten. First flight of the new flying-boat, designated XPB2Y-1 by the US Navy, was on 17 December 1937. It was powered by four 1050 hp Pratt & Whitney XR-1830-72 Twin Wasp engines, and had an armament of two .05-in in the nose and tail, and three .30-in guns — two in the waist and one in tunnel. During initial flight tests it had a tall single fin and rudder, but later a twin tail unit was fitted and the hull was redesigned.

At this time the majority of US Navy flying boat procurements were concentrated on the PBY Catalina, this delaying somewhat orders for the large PB2Y until the last day of March 1939, when Consolidated received a contract for six PB2Y-2s. The cost of one of these giants was as much as the cost of three PBY Catalinas. They were powered by Pratt & Whitney R-1830-78 engines fitted with two-stage superchargers, an increase in armament of six .50-in guns and had a redesigned and deeper hull. Despite an increase in operating weights, the PB2Y-2, now named the Coronado, achieved 255 mph at 19,000 ft. The first to be delivered went to US Navy patrol squadron VP-13 on 31 December 1940. A second production contract was placed with Consolidated on 19 November 1940, this being the PB2Y-3 model, powered by R-1830-88 engines, with self-sealing fuel tanks and eight .50-in guns. Some were equipped with ASV

Excellent air-to-air photo of a USCG Convair PB2Y-3 Coronado long-range patrol aircraft. Known as the Coronado the type served between 1944/46 and served at San Francisco Air Station. The five Coronados operated were a mixture of the PB2Y-3 and -5 models, all being retired during 1946. *(William T Larkins)*

radar in a fairing just behind the cockpit. A total of 210 PB2Y-3s were built.

The Coronado saw very little operational service. A total of thirty-one for the US Navy were converted to PB2Y-3r transports powered by R-1830-88 engines, having the turrets faired over plus other modifications. Others became PB2Y-5 and PB2Y-5R models when fitted with low-altitude R-1830 engines and an increased fuel capacity. The PB2Y-5H was used by the Coast Guard.

Unfortunately details of the Coronado in use with the Coast Guard are not available, although the USCG aircraft inventory dated 31 October 1944 indicates two were on charge. Possibly these were based at the San Francisco air station. Just one BuNo is available from photos, this being BuNo 7138 PB2Y-5H based at San Francisco.

TECHNICAL DATA

Manufacturer:	Consolidated	Take-off power:	4 × 1200 hp
Span:	115 ft	Designation:	PB2Y-3/5H Coronado
Height:	27 ft 6 in	Type:	Patrol flying-boat
Empty weight:	40,935 lb	Length:	79 ft 3 in
Crew:	10	Wing area:	1780 sq ft
Top speed:	213 mph at 20,000 ft	Gross weight:	68,000 lb
Cruise:	141 mph at 1500 ft	Range:	1490 st miles
Sea-level climb:	440 ft per min	Service ceiling:	20,100 ft
Engine:	Pratt & Whitney R-1830-88		

CONVAIR P4Y-2G PRIVATEER

During 1945 the US Coast Guard took delivery of the first of nine single-fin PB4Y-2G Privateer patrol bombers from the US Navy. Their long-range capability and reliability made them particularly suitable for patrol missions over the ocean. Based at San Francisco air station in California, and at Barbers Point, Hawaii, they spent many long flying hours over the Pacific. They also may have been used on the International Ice Patrol. By 1958 only four were still in service, these being returned to the US Navy for disposal by 1960.

The US Navy in 1942 pressed for a force of Consolidated B-24 Liberator land-based patrol bombers to fly long-range overwater missions against shipping and submarines. An agreement dated 7 July 1942 allowed the US Navy a share of the B-24 production. These were designated PB4Y-1, of which five were eventually used by the US Coast Guard.

The development of a B-24 variant more suited specifically to US Navy requirements began on 3 May 1943 when Convair received instructions to allocate three B-24D aircraft for conversion into XPB4Y-2s. This new model retained the same wing and undercarriage, but had a lengthened fuselage, a tall single fin and rudder, and modified engine nacelles. The engines were Pratt & Whitney R-1830-94s minus turbo-superchargers as the aircraft would spend most of their patrol time at low level altitudes.

Used on long-range patrol missions, the Convair P4Y-2G Privateer served the USCG between 1954/59. A total of nine were used and five were operated from the San Francisco Air Station. Depicted is P4Y-2G '6306' seen over San Francisco Bay during 1952 and having the yellow RESCUE fuselage band. *(USCG)*

On 20 September 1943 the first of three prototypes made its first flight, and in October a production order for 660 PB4Y-2s was placed, with the name Privateer adopted. This order was followed by a second order for 710 aircraft a year later. Deliveries commenced in March 1944, ceasing in October 1945, with end of hostilities reducing the number built to 736. A transport version of the Privateer was produced as the RY-3.

Unfortunately little operational use was logged by the PB4Y-2 during World War II, although one patrol squadron — VP-24 — was equipped with the Privateer armed with an ASM-N-2 Bat anti-shipping glide-bomb with radar homing under each wing and designated PB4Y-2B. Other US Navy squadrons flew the PB4Y-2 for several years after the war, some being designated PB4Y-2S when fitted with anti-submarine radar, or PB4Y-2M for weather reconnaissance.

The nine aircraft acquired by the US Coast Guard were designated PB4Y-2G and were used on Air/Sea Rescue (ASR) duties plus weather reconnaissance duties. All turrets were removed, being replaced by large observation blisters. During 1952 four PB4Y-2Gs were still based at San Francisco air station. After retirement the USCG Privateers were returned to the US Navy, initially for storage, then were sold as fire-bombers, with some of the Coast Guard aircraft going to Hawkins & Powers for that purpose, the odd one still being in use in recent fire bombing seasons.

TECHNICAL DATA

Manufacturer:	Consolidated Vultee	Take-off power:	4 × 1350 hp
Span:	110 ft	Designation:	PB4Y-2G Privateer
Height:	30 ft 1 in	Type:	Patrol bomber
Empty weight:	37,485 lb	Length:	74 ft 7 in
Crew:	11	Wing area:	1048 sq ft
Top speed:	237 mph at 13,750 ft	Gross weight:	65,000 lb
Cruise:	140 mph	Range:	2800 st miles
Sea-level climb:	1090 ft per min	Service ceiling:	20,700 ft
Engine:	Pratt & Whitney R-1830-94S		

USCG No.	BuNo	Date	Location
6300	66300	7 July 1958	San Francisco
6302	66302	23 Oct 1952	San Francisco
6306	66306	23 Oct 1952	San Francisco
6260	66260	23 Oct 1952	San Francisco
9688	59688	23 Oct 1952	San Francisco

CONVAIR HC-131A SAMARITAN

When the US Coast Guard acquired seventeen ex-US Air Force C-131A transports during 1977/78 as interim replacement for the ageing Grumman HU-16 Albatross, and pending delivery and introduction of the new HU-25A Guardian, the type was not exactly a stranger to the service.

During 1958 the records reveal that two Convair R4Y-2 transports were delivered to the Coast Guard, possibly on evaluation of the type. These were two transports that had been ordered for the US Air Force as C-131E 57-2551/2 c/n 481/2 and were later allocated US Navy BuNos 145962/3. They were Convair Model 440-71 manufactured respectively on 11 and 14 November 1957 both being delivered on 6 January 1958. Under a Department of Defense Directive dated 6 July 1962, the designations allocated to the US Navy were standardised with those of the US Air Force variants and the two R4Y-2 transports were redesignated C-131G. Details of the evaluation by the Coast Guard are unfortunately not available.

A total of 274 examples of the Consolidated Vultee Aircraft Corporation Convair 240/340/440 series of twin-engined transports was purchased by the US Air Force in no less than nine principal versions of two basic types for training and transportation. The trainer was based on the Convair 240, being originally designated XAT-29, later changed to XT-29. First flight was from Lindbergh Field, San Diego, on 22 September 1949. A production order for forty-six followed, the first T-29A being delivered to the US Air Force on 24 February 1950. The T-29B followed after its first flight on 30 July 1952 with a total of 105 built, while the T-29C, which first flew on 28 July 1953, totalled 119.

The first transport variant produced for the US Air Force was the Convair C-131A named Samaritan, a version of the Convair 240 delivered in 1954, of which twenty-six were built. It could carry thirty-seven passengers or twenty-seven litters, and had a large loading door. The first was delivered on 1 April 1954. The Convair C-131B was based on the Convair 340, with thirty-six being built. A requirement for a VIP and staff transport resulted in thirty-three Convair C-131D and VC-131Ds being built, of which twenty-seven were to the commercial 340 standard and a further six to commercial 440 standard, with improved soundproofing.

US Marine Corps and US Navy fleet support units received a total of thirty-six Convair R4Y-1 twin-engined transports, similar to the US Air Force C-131D and commercial Cv.340, with 2500 hp Pratt & Whitney RA-2800-52W engines. They were delivered from 1952 onwards. On 6 July 1962 the designation was changed to C-131F.

The US Coast Guard actually acquired eighteen Convair C-131A transports from the US Air Force, these retaining the last

Convair HC-131A Samaritan '5790' of the USCG seen in flight profile from the USCG Air Training Centre, Mobile, Alabama. Seventeen were acquired with four more being held in reserve. The Convair was an interim replacement for the ever faithful Grumman HU-16E Albatross and served between 1976 and 1983. *(AP Photo Library)*

four digits of the USAF serial number as the USCG serial. Fourteen of the aircraft were drawn from the permanent storage facility located at Davis-Monthan AFB in Arizona, the remaining four being transferred from the Air National Guard. In case they were required three extra C-131A aircraft were held in reserve at Davis-Monthan unconverted. the eighteenth aircraft acquired by the Coast Guard remained unmodified and utilised by the training school located at the USCG Aircraft Repair & Supply Center at Elizabeth City, North Carolina.

Plans were initiated for one aircraft per month to be modified after the conversion of a prototype and flight trials were completed. Each aircraft underwent a depot maintenance overhaul at Hayes International, Dothan, Alabama. On completion they were flown to the ARSC at Elizabeth City for various avionic and structural modifications which enabled the aircraft to be utilised as efficient Search & Rescue vehicles.

The avionic modification included the rewiring and installation of the following systems: AN/ARA 25 UHF/VHF (AM-FM) DF; AN/ARC 84 VHF transceiver; AN/ARC 94 HF transceiver; AN/ARC 160 VHF-FM transceiver; AN-ARN 44 LF ADF receiver; AN/APM 171 radio altimeter; AN/APN 195 radar and the new ADL-81 LORAN C receiver.

Structural modifications included the installation of a drop hatch, a radioman-navigation position, and several antennas in conjunction with the new avionics installed. Two search observer positions were fitted in the fuselage, an underwater accoustic locator beacon, known as Pinger was fitted, plus a mount for the airborne radiation thermometry (ART) sensor. The instrument panels in the cockpit were standardised, and the cargo compartment was reconfigured.

As the new USCG Convair HC-131A aircraft were completed after modification they were ferried to the Coast Guard Aviation

Training Center at Mobile, Alabama. After initial crew training which included ground personnel, the aircraft were assigned to the following air stations: seven aircraft to Miami air station to be based at Opa Locka Airport, Florida; five to Corpus Christi, Texas; two to San Francisco, California; and two to remain at the ATC at Mobile, Alabama. By 7 July 1977 a total of eight HC-131A Samaritans had been completely overhauled and fitted out. They were 5781, 5782, 5783, 5785, 5786, 5787, 5790 and 5792.

The type was graciously retired during the early 1980s and several ex-USCG HC-131A aircraft are now on the US Civil Register, an example being N27232 ex-5785 seen at Ford Lauderdale, Florida, during October 1982. There was only one accident recorded, this involving 5786 which was written off when it made a hard impact with the runway at Corpus Christi on 18 January 1982.

TECHNICAL DATA

Manufacturer:	Convair	Take-off power:	2 × 2500 hp
Other designations:	Convair 240	Designation:	HC-131A Samaritan
Span:	105 ft 4 in	Type:	Medium range surveillance
Height:	28 ft 2 in		
Empty weight:	29,248 lb	Length:	79 ft 2 in
Top speed:	275 mph	Wing area:	920 sq ft
Cruise:	250 mph	Gross weight:	47,000 lb
Sea-level climb:	1410 ft per min	Range:	450 miles
Engine:	Pratt & Whitney R-2800-99W	Service ceiling:	24,500 ft

CONVAIR HC-131A-CO SAMARITAN

USCG No.	USAF	Designation	Station as of 1 April, 1980
5781	52-5781	C-131A-CO	Corpus Christi, Texas
5782	52-5782	C-131A-CO	Miami, Florida
5783	52-5783	C-131A-CO	
5784	52-5784	C-131A-CO	Miami, Florida
5785	52-5785	C-131A-CO	Miami, Florida
5786	52-5786	C-131A-CO	Mobile, Alabama
5787	52-5787	C-131A-CO	Mobile, Alabama
5788	52-5788	C-131A-CO	Mobile, Alabama
5790	52-5790	C-131A-CO	Mobile, Alabama
	52-5791	C-131A-CO	Davis-Monthan AFB, Arizona — Reserve
5792	52-5792	C-131A-CO	Traverse City, Michigan
5793	52-5793	C-131A-CO	Traverse City, Michigan
5794	52-5794	C-131A-CO	Miami, Florida
5795	52-5795	C-131A-CO	Miami, Florida
	52-5796	C-131A-CO	Davis-Monthan AFB, Arizona — Reserve
	52-5798	C-131A-CO	Davis-Monthan AFB, Arizona — Reserve
5799	52-5799	C-131A-CO	Traverse City, Michigan
5800	52-5800	C-131A-CO	Corpus Christi, Texas
5801	52-5801	C-131A-CO	Corpus Christi, Texas
	52-5804	C-131A-CO	Elizabeth City, North Carolina — Training
5806	52-5806	C-131A-CO	Corpus Christi, Texas

CURTISS MF BOAT

The Curtiss and US Coast Guard records reveal that between 1920 and 1926 several Curtiss MF training flying boats, an updated version of the earlier Curtiss F Boat, were used by the Coast Guard Aviation Group in the development of their ideas of the practicability of using an aircraft to spot derelicts, missing vessels, patrol beaches, provide medical assistance and extend the eyes of the ubiquitous US Coast Guard cutter fleet. The Curtiss MF Boat cruised at 47 knots, had a range of 327 miles and a service ceiling of 5000 feet.

Another historical source reveals that both Curtiss HS-2L and MF flying boats were taken over from the US Navy at the air station located at Morehead City after World War I. The US National Archives indicate that four Curtiss HS-2Ls, one Curtiss MF and one Vought UO-1 were returned to the US Navy during 1925. Coast Guard records indicate an MF Boat was returned to the US Navy by the Coast Guard during February 1926.

The Curtiss MF flying boat was an improved model produced in 1918 and intended to replace the venerable Curtiss F model,

Curtiss F-boat A-2332 tying. The Model F of 1912 became the standard US Navy flying boat trainer and remained in production until 1918. It was replaced by the MF boat (Modified F), an improved model. Several were used by the early Coast Guard Aviation Group. *(Peter M Bowers)*

built in 1912. The MF designation in fact stood for 'Modified F', although apparently there was no detailed resemblance other than the general configuration of a wooden-hulled pusher flying boat. The MF used a flat-sided hull with additional forward buoyancy provided by sponsons added to the side. Initial order by the US Navy from Curtiss was for six aircraft, this being followed by a production order for forty-seven. However the Armistice intervened, and only the first sixteen were delivered. After World War I an additional eighty were produced by the Naval Aircraft Factory.

Initial power-plant was the 100 hp Curtiss OXX-3 although the aircraft used by the US Coast Guard is quoted as having a Curtiss OXX-6 engine. Curtiss built the MF two-seat training flying boat at Garden City, Long Island, and prices quoted were $5821 for those produced by the Naval Aircraft Factory and $3771 for the last twenty produced. Coast Guard records quote $4700.

The Serial number(s) of Curtiss MF flying boat(s) used by the US Coast Guard are not known, likewise the exact commissioning date, but one was decommissioned during February 1926, and returned to the US Navy.

TECHNICAL DATA

Manufacturer:	Curtiss	Prop diameter:	8 ft 1 in
Other designations:	Model 18	Designation:	MF
Span:	49 ft 9 in	Type:	Training flying-boat
Height:	11 ft 9 in	Unit cost:	$4700
Fuel:	40 galls	Length:	28 ft 10 in
Empty weight:	1800 lb	Wing area:	402 sq ft
Crew:	2	Oil:	3 galls
Top speed:	69 mph	Gross weight:	2480 lb
Cruise:	55 mph	Stall speed:	40 mph
Sea-level climb:	350 ft per min	Range:	325 miles
Engine:	Curtiss OXX-6	Service ceiling:	5000 ft
Take-off power:	1 × 100 hp	Two prop blades:	Wood

CURTISS R-6

The Curtiss R-6 floatplane was flown by US Coast Guard aviators during World War I, both during their flight training and later while based overseas in Europe with the cruiser USS *Huntington*. After the war it was used by the fledgling air arm of the USCG Rescue Service to develop the concept of overwater searches. It had a cruising speed of 65 knots, a range of 300 miles and a service ceiling of 4200 feet.

Widely used during 1915-18 was the Curtiss R-series by the US Army, the US Navy and the Royal Naval Air Service for scouting, observation and training. The US Navy models were twin-float seaplanes originally powered the 150 hp Curtiss V-X engine. It was common practice at that time for the pilot to occupy the rear of the two cockpits, with the observer in front. The pilot's vision was somewhat restricted by the wings.

The R, the R-2 and R-2A were all landplanes. Two R-3s were delivered to the US Navy in 1916 fitted with twin floats and powered by 160 hp Curtiss V-X engines. It was a seaplane version of the R-2 with extended wings to carry the added

Curtiss R-6L A-341 one of forty US Navy R-6s converted to R-6L in 1918 by the installation of 360 hp low-compression Liberty engines. The type served in France during World War I and was possibly flown by early US Coast Guard aviators. *(Peter M Bowers)*

weight of the twin-floats. The R-4, R-4L and R-4LM were also landplanes. There does not appear to have been a R-5.

During early 1917 the R-6 appeared, this being a long-wing seaplane like the R-3, differed mainly in having a more powerful 200 hp V-2-3 engine and three degrees of dihedral on the outer wing panels. One R-6, serial A193, was fitted with a single float, the other twenty-five produced, delivered to the US Navy, having twin-floats. The US Army ordered eighteen, most being possibly released to the US Navy before acceptance by the US Army. Those delivered to the US Army were both landplanes and the seaplane version. Curtiss R-6 seaplanes of the US Navy became the first US-built aircraft to serve with the US forces overseas in World War I, when a squadron was assigned on patrol duties in the Azores, based at Ponta Delgada, during January 1918. Average cost of the R-6 was quoted as $15,200. The type served in France on board the cruiser USS *Huntington* which had two US Coast Guard pilots assigned. There is no confirmation that they flew the R-6 though.

It is worth recalling that the US Coast Guard had eighteen fully trained pilots and an aviation engineer officer absorbed on active duty in Europe during World War I.

TECHNICAL DATA

Manufacturer:	Curtiss	Type:	Observation, scouting, training
Span:	571ft 1¼ in	Unit cost:	$15,200
Height:	14 ft 2 in	Length:	33 ft 5 in
Fuel:	76 galls	Wing area:	613 sq ft
Empty weight:	3047 lb	Oil:	8 galls
Crew:	2	Gross weight:	3942 lb
Top speed:	82 mph	Stall speed:	52 mph
Cruise:	75 mph	Range:	300 miles
Sea-level climb:	400 ft per min	Service ceiling:	4200 ft
Engine:	Curtiss V2	Two blade prop blades:	Wood
Take-off power:	1 × 200 hp		
Designation:	R-6		

CURTISS HS-2L

A small number of Curtiss HS-2L flying-boats were acquired by the US Coast Guard from the US Navy in 1920. One source indicates six aircraft, another says four. These were stationed at Morehead City, North Carolina, and they were returned to the US Navy in 1926. The date of commissioning is not known, but all were decommissioned in June 1926. They patrolled the sea lanes off Virgina and North Carolina, assisting in the location of vessels in distress. The three-place flying-boats were powered by a single Liberty twelve-cylinder engine, had a range of 575 miles, a cruising speed of 69 knots and a service ceiling of 5000 feet.

During mid-1917, the Curtiss Aeroplane & Motor Co. Inc. of Garden City, Long Island, and Buffalo, New York, converted the three-seat H-14 twin pusher flying-boat into a single-engined model and assigned the new designation HS = Model H, Single-engine. With the United States involved in World War I, the US Navy ordered a modified version of the HS into large scale production. The existing factories were overloaded so Curtiss could not meet the requirements ordered by the US Navy for the HS. No less than five other manufacturers were given US Navy contracts to build the HS boats under licence from the parent company. Of the 1092 produced, Curtiss built 675. Production ceased when contracts were cancelled after the Armistice and only those aircraft in an advanced stage of construction were completed in 1919. The type remained in US Navy service until 1928.

The HS-1 had a 200 hp Curtiss V-X-X watercooled V-8 engine. On 21 October 1917 the HS-1 was used as the test-bed for the new twelve-cylinder Liberty engine, developing 375 hp in its original form, and in later versions increased to 420 hp. This engine was destined to become the major United States contribution to World War I, and became recognised as one of the world's great aircraft engines.

The original Curtiss order was for 664 flying-boats. The Standard Aircraft Corporation of Elizabeth, New Jersey, was given an order for 250, of which fifty were cancelled. Lowe, Willard and Fowler of College Point, Long Island, was given an order for 200 but fifty were cancelled, the Gallauder Aircraft Corporation of East Greenwich, Connecticut, built sixty and the Boeing Airplane Company of Seattle, Washington, built twenty-five from an original order for fifty. A company located at Santa Barbara, California, the Loughead Aircraft Corporation — known today as Lockheed — built just two.

Fitted with the Liberty engine the HS-1 was redesignated HS-1L. The 180lb depth charges carried by the HS-1L proved to be ineffective against submarines, but the heavier 230lb depth charges could not be carried by the HS-1L. This was overcome by an old Curtiss skill of increasing the wing span. A new centre-

Rare photo taken during 1926 depicting the launch of a US Coast Guard Curtiss HS-2L flying-boat at Morehead City, North Carolina. All of the type were decommissioned and returned to the US Navy during June 1926 *(National Archives No 26-G-20810 via Stephen Harding)*

section twelve ft wider was fitted and one six ft panel was installed between each lower wing panel and the hull. This became the long-span HS-2L. The tail was also enlarged in the vertical, and balance area was added to the rudder. Average cost per unit was $30,000. At least nineteen of the 182 HS boats delivered to the US Navy in France during World War I were the HS-2L model.

Of the four known HS-2L flying-boats operated by the US Coast Guard, two — A1170 and A1240 — were manufactured by Lowe, Willard and Fowler; one — A1474, was manufactured by Standard and one — A1735 by Curtiss, the parent company. The type remained the standard single-engine patrol and training flying-boat in the post-war years, examples remaining in US Navy and US Coast Guard inventory until 1926.

It is left up to the individual as to whether or not these Curtiss HS-2L flying-boats, borrowed from the US Navy, can lay claim as the first US Coast Guard aircraft.

During World War I one US Coast Guard aviator commanded the US Naval Air Station at Chatham, Massachusetts. He piloted one of two Curtiss HS-1 seaplanes which bombed and machine-gunned a German U-boat off the coast of New England. Unfortunately the bombs failed to explode and the enemy submarine escaped.

TECHNICAL DATA

Manufacturer:	Curtiss	Type:	Patrol flying-boat
Span:	74 ft	Length:	38 ft 6 in
Height:	14 ft 7 in	Wing area:	803 sq ft
Fuel:	141 galls	Oil:	13 galls
Empty weight:	4300 lb	Gross weight:	6400 lb
Crew:	3	Stall speed:	55 mph
Top speed:	91 mph	Range:	575 miles
Cruise:	75 mph	Service ceiling:	5000 ft
Sea-level climb:	220 ft per min	Armanent:	One flexible .30-in Lewis gun, two x 230 lb bombs
Engine:	Liberty 12		
Take-off power:	1 × 300 bhp		
Designation:	HS-2L	Two prop blades:	Wood

SERIAL INFORMATION

US Navy BuNo	Commissioned	Decommissioned	
A1170	Date not known	June 1926	Returned to US Navy
A1240	Date not known	June 1926	Returned to US Navy
A1474	Date not known	June 1926	Returned to US Navy
A1735	Date not known	June 1926	Returned to US Navy

CURTISS SOC-4 SEAGULL

The XO3C-1 Seagull was designed to a US Navy specification, being in competition with equivalent Vought and Douglas designs during the early 1930s. Its features were relatively new: full-span leading-edge slats and trailing-edge flaps on the upper wing, a fully enclosed cockpit and a cowling around the Pratt & Whitney radial engine as specified by the US Navy. It was so successful that it survived until the end of World War II, and outlasted two types produced to replace it. This was the second use of the name Seagull by Curtiss, this becoming official when the US government adopted names for popular identification of military aircraft during October 1941.

Primary mission for the new SOC- ordered in prototype form as the XO3C-1 on 19 June 1933 was scout-observation from battleships and cruisers. This required it to be a floatplane, catapulted for take-off and recovered by winch after alighting on the sea alongside the parent ship. It had folding wings with full-span slots and flaps on the top wing and a fully enclosed cockpit for the pilot and gunner. First flown during April 1934, the XO3C-1 had an amphibious landing gear, with two wheels incorporated in the centre main float. This feature was abandoned later, and production models were seaplanes with an alternative under-carriage for land operations.

Evaluation in competition with the Douglas XO2D-1 and Vought XO5U-1 were successful, and the Curtiss biplane went into production as the SOC-1 Seagull. During 1934 there had been a change in US Navy designations by the combination of the scouting and observation roles. Previously, observation types had been deployed on battleships, while scouting types were attached to cruisers. Powered by a Pratt & Whitney R-1340-18 engine, deliveries of the Seagull commenced on 12 November 1935.

The SOC-1 production totalled 135, followed by 40 SOC-2 land-based aircraft which had minor improvements and R-1340-22 engines, plus 83 SOC-3 which were similar to the SOC-2 but had interchangeable alighting gear. A number were equipped with arrester gear for carrier operations, these being designated SOC-2A and SOC-3A after they were modified during 1942. It was US Navy policy to manufacture ten per cent of its own aircraft, resulting in 64 Seagulls being ordered by the Naval Aircraft Factory. Being equivalent to the SOC-3, they were designated SON-1, and SON-1A when fitted with arrester gear.

The three final SOC Seagulls built by the Curtiss Wright Corporation, Curtiss Airplane Division, at Buffalo, New York, were for the Coast Guard and designated SOC-4. During 1942 these went to the US Navy and were modified to SOC-3A standard. Production was complete by the spring of 1938, and when a replacement proved unsatisfactory in operation and was withdrawn in 1944, all remaining SOC Seagulls were hurriedly

This Curtiss SOC-4 Seagull 'V173' is depicted in flight in Washington, not far from its base at Port Angeles Air Station. It was taken on charge during March 1938 and during 1942, along with the remaining two of the type, was transferred to the US Navy with BuNo 48245. *(Gordon S Williams)*

restored to operational status, continuing in service until the end of hostilities in 1945.

Three Curtiss SOC-4 (Model 71F) were accepted by the Coast Guard during 1938, these being improved SOC-3s modified to Coast Guard requirements and assigned USCG serial numbers V171, V172, V173. The first Seagull — V171 — was attached to the USCG cutter *Bibb*, while V173 was based at the USCG air station Port Angeles in Washington state. During 1942 the three aircraft were taken over by the US Navy, modified as mentioned earlier to SOC-3A standard and allocated the US Navy BuNos 48243, 48244, 48245; they had Curtiss c/ns 12412, 12413 and 12414.

These beautiful two-place float biplanes exemplified the final phase of twin-wing days in the Coast Guard, and along with the Grumman JF-2 Duck brought an epoch to its conclusion. Curtiss SOC-4 V171 was operated a landplane version for a while during 1941.

TECHNICAL DATA

Manufacturer:	Curtiss	Type:	Scout and observation seaplane
Contract No.:	Tcg. 27787		
Span:	36 ft	Unit cost:	$48,603
Height:	13 ft 2 in	Length:	31 ft 8 in
Fuel:	170 galls	Wing area:	342 sq ft
Empty weight:	3636 lb	Oil:	12 galls
Crew:	2	Gross weight:	5280 lb
Top speed:	165 mph at 5000 ft	Range:	891 miles
Cruise:	133 mph	Service ceiling:	14,900 ft
Sea-level climb:	880 ft per min	Armament:	One .30 in gun wing wing
Engine:	Pratt & Whitney Wasp R-1340-18		One .30 in gun rear cockpit
Take-off power:	1 × 550 hp		Two 325 lb bombs
Prop diameter:	9 ft	Two-blade prop blades:	Curtiss-fixed pitch.
Designation:	SOC-4 Seagull		

SERIAL INFORMATION

USCG No	Commissioned	Decommissioned
V171	Mar 1938	1942 to US Navy BuNo 48243
V172	Apr 1938	1942 to US Navy BuNo 48244
V173	Mar 1938	1942 to US Navy BuNo 48245

CURTISS SO3C-1/2 SEAMEW

The Curtiss Model 82 SO3C was intended as a replacement for the obsolescent SOC Seagull. It was the third Curtiss model to be named Seagull, although the British Lend-Lease name, Seamew, was subsequently adopted. It was an entirely mid-wing monoplane design, operable either on wheels or floats and conformed to a US Navy requirement drafted in 1937 for a high-speed scouting monoplane. Along with Chance-Vought the Curtiss Wright Corporation, Curtiss Airplane Division of Buffalo, New York, were awarded prototype contracts, producing similar types.

On 6 October 1939 the XSO3C-1 made its first flight. It was a slender, mid-wing monoplane with a low aspect ratio wing and, in its basic configuration, a large central float and underwing stabilisers. Test flights both in this form and as a landplane with a fixed tail landing gear showed serious stability and control problems, so upturned wingtips and an enlarged tail unit were introduced. In addition to being plagued by aerodynamic problems, there was trouble with the inverted air-cooled Ranger V-770 engine, plus a problem of excessive weight caused by additional US Navy equipment requirements.

Deliveries of SO3C-1s in July 1942 with underwing racks for

Curtiss SO3C-1 Seamew landplane of the USCG seen at Port Angeles Air Station, Washington state. Between 1943/44 no less than forty-eight of these scout aircraft were delivered to the service. All but a handful retained their floats, and by July 1944 all had been retired from the Coast Guard. *(Gordon S Williams via W T Larkins)*

two 100lb bombs of two 325lb depth charges. Following production of 141, Curtiss switched to the SO3C-2 version, with arrester gear and provision for carrier operations, plus a bomb rack under the fuselage of the landplane version for a 500lb bomb. Production totalled 459. The SO3C-3 introduced during late 1943 had a more powerful V-770-8 engine and a small reduction in weight in an attempt to improve the overall performance, but production ceased in January 1944, with only thirty-nine built.

In March 1944, the type was retired, some being actually replaced in the US Navy by the earlier SOC Seagull, the aircraft the Seamew had been designed to replace. During 1943 the first batch of forty-five Curtiss SO3C-3 landplanes were delivered to the Coast Guard. The records indicate that three SO3C-1 Seamews were also delivered to the USCG. The Seamew was used on anti-submarine warfare (ASW) patrols off the Gulf and Atlantic coastal waters during World War II. The landplane version was operated by the Coast Guard at Port Angeles in Washington state.

By July 1944 all had been retired from the Coast Guard inventory, including the three early model SO3C-1 aircraft. Many were delivered to the service with wheel landing gear. As they were received, the Coast Guard fitted them as required with float landing gear. Basically built as a scout observation aircraft the SO3C was used mainly for anti-submarine warfare duties.

TECHNICAL DATA

Manufacturer:	Curtiss	Wing area:	293 sq ft
Span:	38 ft	Gross weight:	7000 lb
Height:	14 ft 2 in	Range:	1150 st miles
Empty weight:	4800 lb	Service ceiling:	15,800 ft
Crew:	2	Armament:	One .30 in machine gun fixed
Top speed:	172 mph at 8100 ft		One .50 in machine gun flexible
Cruise:	125 mph		Two 100 lb bombs or 325 lb depth charges
Sea-level climb:	720 ft per min		100 lb or 500b bomb or 325 lb weapon under fuselage
Engine:	Ranger V-770-8		
Take-off power:	1 × 600 hp		
Designation:	SOC3-1/3 Seamew		
Type:	Scouting and observation landplane		
Length:	35 ft 8 in		

CURTISS R5C-1 COMMANDO

During 1943, ten Curtiss R5C-1 Commando transports were acquired by the US Coast Guard and were based at Elizabeth City, North Carolina, to support the huge Aircraft Repair and Supply Center. They were used on logistic support, carrying personnel and cargo to air stations and other USCG units within the United States. Special long-range fuselage fuel tanks were fitted when the transports were required for TDY — temporary duty overseas.

It was as early as 1935 that design studies for a new era transport were commenced with the Curtiss Wright Corporation at St Louis, Missouri. The design was completed in 1936 becoming a 36-passenger all-metal pressurised airliner with room for an additional 8200lb of cargo. The outstanding feature of the design, known as CW-20, was its size for a twin-engined transport, having a wing span of just over 108 feet.

Widely used in the Pacific theatre of operations during World War II, the Curtiss C-46 Commando was the largest and heaviest twin engined aircraft in operational service with the US Army Air Force. The basic design became militarised, serving all theatres with both the USAAF and the US Marine Corps. It continued with service in the Korean conflict and even operated during the early days of the Vietnam war.

Early military Commandos suffered from a high degree of mechanical problems, not related to the aircraft design. Minimum maintenance plus operating in harsh environments were unforeseen by the C-46 designers. With a greater load-carrying capacity and having better performance at high altitudes

Ten Curtiss R5C-1 twin-engined Commando transports were acquired by the USCG during 1943, serving until 1950. They were operated on logistic support missions, carrying personnel and cargo to USCG units in the US and overseas. Unfortunately very little information is available on the use of this transport with the Coast Guard. *(AP Photo Library)*

compared with the Douglas C-47 they were assigned to the Pacific theatre and the CBI — China, Burma, India theatre where they contributed to the success of the supply operation over the Hump supplying material to China from India after the capture of Burma by the Japanese. Subsequent improvements made the C-46 one of the most reliable aircraft of World War II. Not counting the prototype, a total of 3182 C-46 and R5C-1 Commandos were built and delivered between 1942 and 1945.

The first Curtiss C-46 Commando was rolled out of the Curtiss factory at Buffalo, New York, in May 1942 and delivered to the USAAF in July. After a successful introduction into service, coupled with a pressing need to provide an airlift capability for the US Army, orders were rapidly multiplied. Equipped with a large cargo loading door in the rear fuselage, a heavy steel cargo floor and folding seats along the cabin sides for forty fully equipped troops, the C-46A followed the small number of twenty-five C-46s, the company building 1041 of the C-46As at Buffalo. Production also took place at Louisville and St Louis, where 439 and ten were built respectively. Higgins of New Orleans received a contract to construct 500 but only two were completed. The C-46A was powered by 2000hp R-2800-51 engines and gross weight was 49,600lb.

The C-46 Commando saw limited service in the European theatre of operations serving with US Ninth Troop Carrier Command units for glider towing and dropping paratroops. The first C-46A-CK was modified to XC-46B with 2100hp R-2800-34W engines and a new stepped windscreen, but this variant was not adopted. The C-46D was the next major production variant, retaining the R-2800-51 engines, but having a revised nose and a troop door in the starboard side. The Curtiss-Wright Corporation, Airplane Division at Buffalo built 1400 followed by 234 C-46Fs which were similar but had R-2800-75 engines and blunt wing tips. The St Louis plant built seventeen C-46Es, similar to the 'F' model but having a single cargo-loading door and a stepped windscreen like the XC-46B.

Only one C-46G was built at Buffalo, it having two doors and blunt wing tips and powered by R-2800-34 engines as the contract for 500 was halted by VJ-Day. Also cancelled was the C-46H with more powerful engines and twin tail wheels. The XC-46K was to have had 2500hp Wright R-3350 Double Cyclone engines, but these were used to power three XC-46L transports.

Of the 3182 C-46 transports produced at the three plants, 160 went to the US Marine Corps, who used them primarily to support the campaign against the Japanese in the Pacific theatre of operations. They were powered by Pratt & Whitney R-2800 engines, carried fifty troops and were equivalent to the US Army Air Force C-46A model.

The Curtiss R5C-1 Commando transports operated by the US Coast Guard were finally retired in 1950. A USCG aircraft inventory dated 1 May 1947 does not include any R5C-1 transports, but the inventory for January 1950 includes two. A photograph of a general muster at Elizabeth City on 26 May 1950 shows two R5C-1 transports parked. Only one R5C-1 Commando used by the USCG can be identified, this being BuNo 39537 which was struck off charge from the US Navy on 31 May 1947 as surplus. It was reinstated on 31 July 1948 and assigned to the US Coast Guard. It was held by the US Navy at NAS Norfolk, Virginia.

TECHNICAL DATA

Manufacturer:	Curtiss	Take-off power:	2 × 2000 hp
Other designations:	C-46	Designation:	R5C-1 Commando
Span:	108 ft 1 in	Type:	Troop and freight transport
Height:	21 ft 9 in		
Empty weight:	32,400 lb	Length:	76 ft 4 in
Crew:	4	Wing area:	1360 sq ft
Top speed:	269 mph at 15,000 ft	Gross weight:	56,000 lb
Cruise:	183 mph	Passengers:	50 or 10,000 lb cargo
Sea-level climb:	1300 ft per min	Service ceiling:	27,600 ft
Engine:	Pratt & Whitney R-2800-51		

DASSAULT-BREGUET HU-25A/B GUARDIAN HU-25C INTERCEPTOR

Early in 1977, on 5 January, a news release from the Falcon Jet Corporation of Teterboro, New Jersey, showed a scale model of the new Coast Guard HX-XX aircraft, the Falcon 20G. Award of the $205 million US government contract to the Falcon Jet Corporation to supply the US Coast Guard with forty-one medium range surveillance jet aircraft was announced in Washington, DC, by Secretary of Transportation, William T Coleman Jr and Coast Guard Commandant, Admiral Owen W Siler. The HX-XX programme forecast a $84 million payroll affecting such key industrial areas as southern California, Arizona, Arkansas, New Jersey and Iowa. Final assembly of all structural components including engines and avionics were to be shipped to the Falcon Jet's Little Rock Division facility in Little Rock, Arkansas. This news release coincided with a similar release from the USCG No 1-77.

It was during 1967, that the USCG commenced a search for a suitable replacement for its ageing fleet of Grumman HU-16E Albatross amphibians, at which time more than seventy of the 'Goat' remained in service. What followed was a mammoth and expensive evaluation exercise during which more than thirty aircraft types were studied informally. A volume could be written on the evaluation which the Coast Guard carried out in the ensuing ten years, prior to a final decision on the selection of a new medium range surveillance (MRS) aircraft. By the early 1970s the requirement for a new MRS aircraft became more clearly established. The choice between a turboprop, turbojet and turbofan type became less certain, and operational six month evaluation of a Israeli Aircraft Industry Westwind 1123 and a Cessna Citation in USCG livery took place. A Grumman Gulfstream I was also included in the evaluation programme.

In late 1974, a decision was made to proceed with the issuance of a Military Interservice Procurement Request (MIPR) to the US Navy for the purchase of the Rockwell Sabre 75A. It was an abortive attempt for several reasons. A second Request for Technical Proposals (RFTP) was issued on 19 January 1976, with five US manufacturers responding to this second solicitation, most responding with two or more proposals.

The choice eventually fell upon a version of the French built and designed Dassault-Breguet Mystere 20, marketed in the US by Falcon Jet Corporation as the Falcon 20 business jet. The Coast Guard's version was the Falcon 20G designated HU-25A Guardian by the US Department of Defense, although never procured by any other US military service. The first aircraft to appear in USCG livery carried the US civil registration N1045F.

An initial Coast Guard programme extended over a seventy-one-month period with the first aircraft scheduled for delivery in July 1979, thirty months after the award of contract. Subsequent

Forty-one HU-25 Guardian aircraft were procured for the USCG, the type being involved in the drug interdiction operations. Depicted is HU-25A Guardian '2116' from Cape Cod USCG Air Station, Massachusetts. The type entered service in 1982 and the civil version is known as the Falcon. *(AP Photo Library)*

deliveries were scheduled at the rate of one per month for forty-one months. The aircraft was fabricated at various facilities in France, final assembly taking place at Merignac, near Bordeaux. The aircraft was then crated in a partially disassembled state and flown by stretched Lockheed C-130 Hercules to Little Rock, Arkansas, for assembly. Final assembly included the fitting of Garrett ATT-3-6-2C engines, an integrated Collins avionics navigation and communications package, a Texas Instruments APS-127 radar, Loran-C receiver, other Comm-nav equipment, interior panels and furnishings, hard points and console to accommodate and manage a wide variety of surveillance sensor systems, a drop hatch for air delivery equipment to distressed vessels and cabin search windows.

In terms of dollars the HU-25A procurement was one of the largest ever undertaken by the Coast Guard. The Falcon Jet bid for the contract, one of the lowest, was $204,846,291.

The HU-25A Guardian was fitted with search windows on either side, four hardpoints under the fuselage to carry supply or rescue packs for four wing hardpoints for sensor pods or other loads. Normal crew consisted of two pilots, a surveillance systems operator and two search crew members.

First flight of a development HU-25A aircraft took place on 28 November 1977 and certification was obtained on 21 June 1981. Deliveries were initiated on 19 February 1982 and completed on 8 December 1983 the new Guardian eventually entering service at nine Coast Guard air stations. At a later date, seven Guardians were modified as HU-25Bs, carrying a SLAR — sideways locking radar — pod on the fuselage, and an APS-131 IR/IV line scanner under one wing, for law enforcement missions involving the detection and identification of ships causing oil spills. All are configured with the AIREYE system which was originally evaluated on a HU-16E Albatross.

HU-25C INTERCEPTOR

As part of its effort to interdict airborne drug smugglers the Coast Guard modified nine Guardian jets during 1988, designated HU-25C Interceptor. The modifications were funded by the US Anti-Drug Abuse Act of 1986 and approved by the National Drug Policy Board, which clarified the Coast Guard's role in air interdiction.

The Interceptor's new equipment included a Westinghouse AN/APG-66 long-range search radar found in the General Dynamics F-16 Fighting Falcon fighter, and a WF-360 forward looking infra-red (FLIR) sensor in a radome under the fuselage. This provides all-weather day/night capability to intercept, classify and track suspect maritime and airborne targets.

It was also equipped with secure communications capabilities in the HF, UHF and VHF-FM frequency ranges. It can communicate with virtually all other military and civilian law enforcement

All the HU-25 Guardian fleet are equipped with hard points and modified and updated to include a computer, and surveillance pods on the wings. These include infra-red/ultra-violet (UVLS), line scanners and sideways looking radar (SLAR). Depicted is '2125' marked *Sacramento* and equipped with SLAR and underwing pod. *(AP Photo Library)*

agencies in a secure or non-secure mode. The addition of these systems to the high dash speed and excellent slow flight characteristics makes the modified HU-25 a state of the art aircraft that will enable the Coast Guard to more effectively engage the smugglers and drug barons in the maritime area. The fuel capacity consists of 9870lb of JP-4, and 10,431lb of JP-5/JP-8. Having a Mach 0.85 speed capability, the aircraft can quickly be assigned between Coast Guard Districts in response to urgent Search & Rescue (SAR), law enforcement or any other mission. This rapid response capability permitted the Coast Guard to procure a minimal number of aircraft. Included in the aircraft price was a requirement for continued technical support over the guaranteed 30,000 hour life of the aircraft.

Air stations equipped with the HU-25C Interceptor include Miami, Florida, based at Opa Locka and Mobile, Alabama, the latter being the US Coast Guard Aviation Training Center (AVTRACEN). In line with low profile restriction, initiated in 1989, removal of all air station identity has included the versatile HU-25 series of aircraft.

TECHNICAL DATA

Manufacturer:	Falcon Jet Corporation	Take-off power:	2 × 5440 lb thrust at sea-level
Other designations:	HU-25C	Designation:	HU-25A/B Guardian
Span:	54 ft	Type:	Medium-range surveillance
Height:	18 ft	Unit cost:	$4,996,251
Fuel:	1534 US galls	Length:	56 ft
Empty weight:	19,000 lb	Wing area:	450 sq ft
Crew:	Two pilots + 3 or 5 crewmen	Gross weight:	33,510 lb
Top speed:	531 mph at 40,000 ft	Range:	2590 miles
Cruise:	475 mph at 41,000 ft	Service ceiling:	41,000 ft +
Sea-level climb:	5440 lb at 59°F		
Engine:	Garrett ATF3-6-2C turbofans		

TECHNICAL DATA

Manufacturer:	Falcon Jet Corporation	Take-off power:	2 × 5440 lb thrust at sea-level
Other designations:	HU-25A/B	Designation:	HU-25C Interceptor
Span:	54 ft	Type:	Medium-range surveillance
Height:	18 ft	Unit cost:	$4,996,251 +
Fuel:	1534 US galls 9870 lb JP-4 10,431 lb JP5/JP8	Length:	56 ft
Empty weight:	25,500 lb	Wing area:	450 sq ft
Crew:	Two pilots, 3 crewmen	Gross weight:	32,000 lb
Top speed:	350 knots sea level 380 knots at 20,000 ft +	Range:	2045 miles at low altitude
Sea-level climb:	5440 lb at 59°F	Service ceiling:	41,000 ft +
Engine:	Garrett ATF3-6-2C turbofans		

SERIAL INFORMATION

USCG	Commissioned	Decommissioned
2101 to 2141		

DOUGLAS O-38C

Last of one of the longest-lived designs built between the war years was the Douglas biplane series that commenced in 1924 with the XO-2. The O-38 series was the last of the Douglas observation biplanes with a few still in service at the time of the Pearl Harbor attack by Japan.

The batch of forty-six O-38s were similar to the succeeding models, designated O-29 and O-32, except for 525hp Pratt & Whitney R-1690-3 Hornet engines. The first of the series was converted to O-38A, the sixty-three O-38Bs that followed being just improved O-38s. There was a single O-38C serial USAAC 32-394, which was a O-38B procured by the US Army Air Corps for the Coast Guard.

A single Douglas O-38C observation aircraft was ordered by the US War Department for the USCG during 1931. This O-38C 32-394 became registered 'CG-9' and was commissioned in December 1931. It later became 'V108' and was withdrawn from use in April 1934, after being involved in an accident. *(William T Larkins)*

Ordered by the US War Department during late 1931 for use by the Coast Guard, the O-38C 32-394 became registered CG-9 after being commissioned in December 1931. It later became V108 and was identical to the O-38B being powered by a 525hp Pratt & Whitney R-1690-7 radial engine and was fitted with dual controls.

The Coast Guard O-38C was decommissioned during April 1934, after it had been involved in an accident.

TECHNICAL DATA

Manufacturer:	Douglas	Prop diameter:	10 ft 1 in
Contract No.:	Air Corps AC-4553	Designation:	O-38C
Span:	40 ft	Type:	Observation
Height:	10 ft 10 in	Unit cost:	$17,900
Empty weight:	3050 lb	Length:	31 ft 3 in
Crew:	2	Wing area:	360 sq ft
Top speed:	147 mph	Gross weight:	4350 lb
Engine:	Pratt & Whitney Hornet R-1690-7	Service ceiling:	20,700 ft
Take-off power:	1 × 525 hp	Two-blade prop blades:	Metal, fixed pitch

DOUGLAS RD, RD-1/2/4 DOLPHIN

One of the reasons for the confusion that exists over the correct identification of the various Douglas Dolphins used by the US Coast Guard, is that four entirely different models were in service at the same time in the 1930s. Dimensional and performance figures have caused confusion over the years, and even the historian with the Douglas Aircraft Company in California has discovered discrepancies in the company records. The Douglas RD started as the Sinbad, going through some iterations before the final configuration was established. It appeared to be smaller and lighter than the later Dolphins. The US Coast Guard aircraft register for 1933 shows the RD named *Procyon* CG-27 based at Cape May, New Jersey, and allocated the international radio call-sign 'NUMRG' and Coast Guard call-sign '24 G'. Apparently this first production aircraft was delivered to the USCG in New York direct from the Douglas factory in February 1931. This was a flying-boat, not an amphibian. The RD-2 *Adhara* was delivered in July 1932, and in 1933 it was based at Gloucester, Massachusetts, with the international radio call-sign 'NUMRJ' and Coast Guard call-sign '24 J'. The RD-1 *Sirius* followed on 5 August 1932, being based at Miami, Florida, in 1933 with international call-sign 'NUMRH' and Coast Guard call-sign '24 H'. The first RD-4 was not delivered until nearly three years later, on 20 February 1935. All four types were externally different in fuselage, engine and tail configuration.

HISTORY

Donald Wills Douglas was a lover of the sea, being a keen yachtsman, so it was no surprise when ten years after founding

Douglas RD-2 Dolphin '129' *Adhara* seen in the livery adopted by the USCG for its aircraft in the 1930s. Delivered in July 1932 it later became 'V111' but unfortunately crashed in March 1937. A total of thirteen Dolphins were operated by the service involving four variants of this popular high-wing cantilever monoplane. *(USCG)*

the Davis-Douglas Company, he combined his vocation and avocation by building a flying-boat of his own design, named the Sinbad. This was the prototype of the Dolphin whose success can only be contributed to the versions purchased by the military services including the US Coast Guard. The Sinbad was a luxury air yacht, and a victim of the Great Depression when it was completed in July 1930.

The aircraft was a high-wing cantilever monoplane of wood and metal construction, with an all-metal hull and wood covered cantilever wing. The two pilots were accommodated in an enclosed cabin located just forward of the wing, with a cabin for eight passengers within the hull and beneath the wing. A pair of 300hp Wright J-5C Whirlwind radials mounted above and forward of the wing on multiple struts initially faired over to reduce drag. The Sinbad was a flying-boat and was fitted with beaching gear, while the Dolphin production version was an amphibian which was fitted with an undercarriage.

Flight testing of the Sinbad took place from Santa Monica Bay during July 1930. At this time the engines were enclosed in metal covered cowlings which sloped rearward and downward for attachment to the upper surface of the wing. These flight trials revealed the need for a number of major modifications to the airframe and the engines. The latter needed to be mounted higher, and the fairings were replaced by conical nacelles of welded chrome-molybdenum steel which were attached to the upper surface of the wing by six bracing struts. To increase rigidity and smooth out the airflow, a small auxiliary aerofoil between the engine nacelles was added. Other minor modifications included the addition of twin auxiliary tail fins and the relocation of the fixed floats further outboard.

With these modifications complete the Douglas Sinbad flying-boat registered X145Y was purchased for $31,500 by the US Coast Guard. It was delivered to the USCG on 9 March 1931 being initially numbered 27, this being changed to 227 in February 1935, and to V106 in October 1936. It served the Coast Guard until November 1939, being designated RD. The next Douglas RDs for the USCG were true amphibians.

During flight testing of the Sinbad, the Douglas Aircraft Company was planning the production of the Dolphin, an amphibian flying-boat based on the modified Sinbad and intended for both the military and civil market. The production Dolphins were fitted with a retractable undercarriage consisting of a main gear attached to the hull by hinged V-struts, and attached to the undersurface of the wing by an oleo leg. A tailwheel replaced the skid of the Sinbad, being attached to the hull aft of the rear step. While in flight, or during water operations, the main undercarriage was raised by shortening the oleo legs, lifting the wheels above the waterline, while the tailwheel pivoted upward until it was close to the hull.

A total of fifty-nine Dolphins were built between 1931 and

1934, which included not less than seventeen variants or models, as the Dolphins were custom-built for each civil customer or produced in small batches for the US military services including the Coast Guard. There were also improvements, modifications and/or engine changes introduced on the production line at Santa Monica. Initially a single Dolphin, designated RD-2, using the US Navy designation system, was ordered by the US Army Air Corps for the Coast Guard, successively carrying the USCG numbers 29 changing to 129 in January 1935, and V111 with effect from October 1936. This flying-boat was similar to the Air Corps YIC-26, but was powered with two Pratt & Whitney Wasp R-1340-10 engines and had the c/n 1122.

The one and only RD-2 Dolphin commissioned during 1932 was the administrative aircraft for the Secretary of the Treasury until 1937. Except for the engines — two Pratt & Whitney Wasps — it was basically the same as the RD-1.

Commissioned during November 1934 were ten Douglas RD-4s for the Coast Guard, this bringing to a close the production of US military Dolphin variants. They were generally similar to the US Navy RD-3 but were powered by two 420hp Pratt & Whitney Wasp C1 engines with the fuel tankage increased to 252 US gallons or 954 litres. The USCG serials were initially '130' to '139' these being changed to V125 to V134 from 13 October 1936.

Seen on the ramp at San Francisco Air Station on 6 March 1942 is Dolphin RD-4 'V128' *Vega* in World War II camouflage. It was commissioned during February 1935. The Dolphin was possibly one of the most popular of the many unique types operated by the USCG. *(William T Larkins)*

The US Coast Guard aircraft were used extensively in search and rescue (SAR) missions and as flying lifeboats, often flying far out to sea from several air stations to rescue stricken mariners or seamen in need of urgent medical care to hospitals ashore. Upon US entry into World War II in December 1942 the US Coast Guard became part of the US Navy and the surviving Douglas RD-4s were assigned to security patrols along the United States seaboard.

In the 1930s the US Coast Guard was able to make use of one of those fascinating peacetime luxuries that are normally limited to small organisations — the naming of individual aircraft. This practice was so common that many official Coast Guard communications, and nearly all press releases and newspaper stories, referred only to the name of the aircraft. These names appeared on each side of the nose and serve as a very accurate means of identification of the individual aircraft. In photographs where the USCG serial number is not visible, it is often the only means of positive identification. The RD designation stood for Multi-engine Transport, Douglas.

TECHNICAL DATA

Manufacturer:	Douglas	Take-off power:	2 × 300 hp
Military spec no:	SD-178-4	Designation:	RD Dolphin
Contract No.:	Tcg. 12608	Type:	Transport amphibian
Span:	60 ft	Unit cost:	$31,500
Height:	14 ft 7 in	Length:	42 ft
Fuel:	180 galls	Wing area:	575 sq ft
Empty weight:	5605 lb	Oil:	9 galls
Top speed:	136 mph	Gross weight:	8000 lb
Cruise:	108 mph	Stall speed:	60 mph
Sea-level climb:	700 ft per min	Range:	600 miles
Engine:	Wright Whirlwind	Service ceiling:	14,000 ft

TECHNICAL DATA

Manufacturer:	Douglas	Designation:	RD-1 Dolphin
Military spec no:	Spec. SD-178	Type:	Transport amphibian
Contract No.:	Air Corps AC-4460	Unit cost:	$36,500
Span:	60 ft	Length:	42 ft 2 in
Height:	14 ft 1 in	Wing area:	562 sq ft
Fuel:	180 galls	Oil:	18 galls
Empty weight:	6127 lb	Gross weight:	8415 lb
Crew:	2	Passengers:	6
Top speed:	152 mph at 3000 ft	Stall speed:	60 mph
Sea-level climb:	860 ft per min	Range:	466 miles
Engine:	Wright Whirlwind 300, E-965-E	Service ceiling:	17,300 ft
Take-off power:	2 × 435 hp	Two-blade prop blades:	33C1-6, metal
Prop diameter:	8ft 6in		

TECHNICAL DATA

Manufacturer:	Douglas	Designation:	RD-2 Dolphin
Military spec no:	SD-178-2	Type:	Administrative transport
Contract No.:	Air Corps AC-4921	Unit cost:	$43,250
Span:	60 ft	Length:	45 ft 2 in
Height:	14 ft 5 in	Wing area:	562 sq ft
Fuel:	240 galls	Oil:	18 galls
Empty weight:	6969 lb	Gross weight:	9980 lb
Crew:	2	Passengers:	6
Top speed:	162 mph at 6000 ft	Stall speed:	64 mph
Cruise:	105 mph	Range:	770 miles
Sea-level climb:	1020 ft per min	Service ceiling:	14,900 ft
Engine:	Pratt & Whitney Wasp R-1340-10	Two-blade prop blades:	Hamilton Standard 3792F
Take-off power:	2 × 300 hp		
Prop diameter:	9ft		

TECHNICAL DATA

Manufacturer:	Douglas	Take-off power:	2 × 454 hp
Military spec no:	SD-178-4	Designation:	RD-4 Dolphin
Contract No.:	US Navy 34223	Type:	Transport amphibian
Span:	60 ft	Unit cost:	$60,000
Height:	14 ft 7 in	Length:	45 ft 3 in
Fuel:	240 galls	Wing area:	592 sq ft
Empty weight:	6467 lb	Oil:	14 galls
Crew:	3	Gross weight:	9737 lb
Top speed:	147 mph	Passengers:	6
Cruise:	110 mph	Stall speed:	63 mph
Sea-level climb:	710 ft per min	Range:	660 miles
Engine:	Pratt & Whitney Wasp R-1340-CI	Service ceiling:	14,500 ft

DOUGLAS RD, RD-1/2/4 DOLPHIN

USCG Nos.			Name	Type	c/n	Commissioned	Decommissioned	Notes
27	227	V106	*Procyon*	RD	703	9 Mar 1931	Nov 1939	
28	128	V109	*Sirius*	RD-1	1000	5 Aug 1932	Nov 1939	
29	129	V111	*Adhara*	RD-2	1122	Jul 1932	Mar 1937	Crashed
	130	V125	*Spica*	RD-4	1268	Nov 1934	Jan 1943	
	131	V126	*Mizar*	RD-4	1269	Feb 1935	Aug 1941	Crashed
	132	V127	*Alloth*	RD-4	1270	Feb 1935		
	133	V128	*Vega*	RD-4	1271	Feb 1935		
	134	V129	*Deneb*	RD-4	1272	Mar 1935	Jul 1942	
	135	V130	*Aldebaran*	RD-4	1273	Feb 1935	Aug 1935	Crashed
	136	V131	*Rigel*	RD-4	1274	Mar 1935	Jul 1940	
	137	V132	*Capella*	RD-4	1275	Apr 1935	Jun 1943	
	138	V133	*Bellatrix*	RD-4	1276	Apr 1935		
	139	V134	*Canopus*	RD-4	1277	Apr 1935	Aug 1942	

DOUGLAS R4D-5 SKYTRAIN

On 16 September 1940 the US Navy signed an initial contract for thirty Douglas R4D-1 transports, followed eventually by another 103, powered by 1200hp Pratt & Whitney R-1830 Twin Wasp two-row, 14-cylinder radial engines. The Douglas R4D-1 was basically a commercial-standard DC-3 with refinements for military use. There were only two Douglas R4D-2 transports, these being powered by Wright Cyclone R-1830 engines.

This new Douglas R4D- transport had the equivalent counterpart in the US Army Air Force. The Douglas R4D-3 and R4D-4 were ex-airline transports impressed into service with the US Navy. Later variants were the R4D-5, which had a 24-volt electric system similar to the USAAF C-47A Skytrain. The R4D-6 was equivalent to the USAAF C-47B, and the R4D-7 was identical to the USAAF TC-47B navigation trainer.

During May 1943 the first batch of a total of eight Douglas R4D-5 transports were delivered to the US Coast Guard to be employed search and rescue plus logistic support work. The second and final batch of R4D-5 transports were delivered to the service during July1944. By 1956 all but four remained in Coast Guard service, these being gradually phased out over the next few years. Records show that on 16 May 1961 Douglas R4D-5 BuNo 12446 was still held on inventory at the Aircraft Repair & Supply base located at Elizabeth City, North Carolina.

At least one R4D-5 BuNo 17243 was based at Port Angeles air station in Washington state, to provide support for the Alaska LORAN — long range navigation — chain, plus support of the USCG air detachments at Annette Island and Kodiak Island in Alaska. Photos of this Douglas transport are dated March 1950. A large observation blister was fitted to the aft window on the fuselage of some USCG R4D-5s, this being useful when involved in search and rescue missions.

On 28 May 1947 Douglas R4D-5 BuNo 17243 completed overhaul at San Diego air station, California, while BuNo 12446 was a visitor to the San Francisco air station, California, on 30 October 1949. On retirement BuNo 17183 was stored at the huge US Navy facility at Litchfield Park, Arizona, during 1961, being later transferred to the famed boneyard located at Davis-Monthan AFB also in Arizona.

OPERATIONS

The Douglas R4D-5 transports operated by the Coast Guard were already veterans of World War II, flying many hours over and into the many theatres of operations. It is interesting to record a potted history of the three USCG transports identified.

One: c/n 9759 BuNo 12446. Delivered to the US Navy on 24 June 1943, immediately being enrolled in the huge Naval Air Transport Service — NATS — being initially operated by VR-3

Eight Douglas R4D-5 Skytrain transports were operated by the US Coast Guard from 1943 and the last one retired in the 1960s. Depicted parked at the huge Aircraft Repair & Supply Center at Elizabeth City, North Carolina, is Douglas R4D-5 BuNo 12446 on 16 May 1961. *(USCG)*

Squadron based at Kansas City, Kansas. It served consecutively with VR-4 and VR-13 Squadrons before being assigned to Commander Air Pacific — COMAIRPAC. It returned to NATS serving with VR-6 and VR-11 Squadrons, ending up at San Diego on 30 April 1946, possibly with the Coast Guard.

Two: c/n 12798 BuNo 17183. Delivered to the US Navy on 26 March 1944, going immediately to the US Marine Corps and serving with VMR-152 with effect from August 1944. It was assigned to NAS San Diego in April 1946, being struck off charge on 31 July 1946, when presumably it went to the Coast Guard. In 1964 it was civil registered N2204S which it retained until 1984 when it ws sold to Bolivia as CP-1940 so could be still flying.

Three: c/n 25441 BuNo 17243. Delivered to the US Navy on 31 July 1944, its first base being San Diego. Up to 1945 it served with a variety of units going to COMAIRPAC on 7 February 1946, then to NATS with VR-6 Squadron in March 1946, returning to San Diego in April 1946. It then went on the Coast Guard inventory.

TECHNICAL DATA

Manufacturer:	Douglas	Take-off power:	2 × 1200 hp
Other designations:	C-47	Designation:	R4D-5 Skytrain
Span:	95 ft	Type:	Transport
Height:	17 ft	Length:	63 ft 3 in
Empty weight:	16,578 lb	Wing area:	987 sq ft
Crew:	3	Gross weight:	29,000 lb
Top speed:	227 mph at 7500 ft	Passengers:	27
Cruise:	135 mph	Range:	1975 st miles
Sea-level climb:	940 ft per min	Service ceiling:	22,500 ft
Engine:	Pratt & Whitney Twin Wasp R-1830-92		

DOUGLAS R5D-3/4/5 SKYMASTER

During 1958, four-engined Douglas R5D Skymaster transports replaced the ageing US Coast Guard Boeing PB-1G Flying Fortress aircraft on the International Ice Patrol. The archives reveal that six Douglas R5D-3 Skymasters were acquired by the Coast Guard during 1945, and a total of fifteen of these workhorses appear to have served with the service between 1945 and 1962, although only one was still held on inventory in 1965. The type was finally retired, being replaced by the ubiquitous Lockheed C-130 Hercules.

It was in June 1939 that American, Eastern and United Airlines plus the Douglas Aircraft Company of Santa Monica, California, shared the view that there was a need for an aircraft similar in capacity to the four-engined experimental DC-4E completed in May 1938 and evaluated by United Airlines. Performance was not good, maintenance presented problems, and operating economics were very disappointing. The sponsoring airlines agreed with the Douglas company to suspend the DC-4E development in favour of a new, less complex, DC-4 project which was to lead to the military C-54. The Douglas engineering team led by Arthur Raymond and Ed Burton designed a new aircraft which was also designated DC-4. Initial reactions from the airlines were enthusiastic and the orders came in. However, the war clouds in Europe increased military orders from Douglas for the British and French Purchasing Commissions plus the US armed forces, but production of the DC-4 went ahead. The Japanese attack effected once again the production plans of the DC-4. Aircraft already under construction at Santa Monica were taken over by the US Army Air Force, becoming C-54s.

No prototype was built, or in fact necessary, and the first production aircraft was completed in February 1942, making its first flight at Clover Field on 14 February. The success of the C-54 and its trouble-free development trials programme provided the USAAF with a long-range heavy logistic transport, a type urgently needed by the worldwide scale of operations into which the USA had been forced without proper preparation. There was a requirement for both trans-Atlantic and trans-Pacific flights which led to the modification of the first twenty-four aircraft to install auxiliary fuel tanks in the main cabin. This increased the total fuel capacity from 2012 US gallons to 3580 US gallons, but naturally reduced the seating capacity. These early C-54-DO were only a stop-gap version taking advantage of nearly completed commercial DC-4s and the first true military C-54A-DO equipped to carry either troops or cargo.

Quantity production was planned and to support the Santa Monica plant, Douglas established a new production line at its Chicago, Illinois, factory. As a result production increased rapidly, with twenty-four in 1942, seventy-four in 1943, 354 in

1944, 710 in 1945, but only one in 1946 due to cancellation of contracts after VJ-Day. The Chicago plant accounted for some two-thirds of the total production of the type.

The first military model constructed at both Santa Monica and Chicago was the C-54A, now named Skymaster, structurally redesigned for the carriage of heavy cargo with a reinforced floor, a large door and a boom hoist and winch. Troops and cargo could be carried in the fuselage, and provision was made for large items of cargo to be carried on suspension points under the fuselage. Fuel capacity was increased and the engines were Pratt & Whitney R-2000-7. Integral fuel tanks in the wings replaced two fuel cells in the fuselage in the C-54B variant, and cabin fittings were provided for stretchers. Gross weight was increased to 73,000lb with a seating capacity of fifty. One hundred were built at Santa Monica and one hundred and twenty at Chicago.

Most produced variant was the C-54D, similar to the 'B' but with R-2000-11 engines. It was the final model built at Chicago who produced 350 aircraft in 1944/45. Santa Monica produced the C-54E with a combination of seats and cargo fittings, plus extra bag tanks in the wings giving an increase in range. After producing seventy-five C-54Es, the Californian factory built seventy-six C-54G models similar to the 'E' but powered by 1450hp R-2000-9 engines. By this time a total production of 952 had been achieved, plus a further 211 more on contract for the US Navy, some of which later went into US Coast Guard service.

There were other models, such as the experimental XC-54F which appeared only in mock-up form. The C-54H, a similar project with R-2000-9 engines, was cancelled when World War II came to an end. There were also the C-54J, a variant of the G with a new interior, also cancelled, and the XC-54K, long-range development of the E with 1425hp R-1830-HD engines, of which only a single example was built.

Service of the transport was worldwide, and during the course of three years of World War II flying an unequalled safety record was achieved. In making 79,642 ocean crossings up to VJ-Day, only three Douglas C-54 Skymasters were lost. A regular and reliable service across the North Atlantic, averaging twenty round flights a day for months was established. It flew the Pacific from the west coast of the US to the Philippines and Australia, and across the Indian Ocean from Ceylon to Australia, a 3100 mile flight leg. It served in North Africa, in Alaska, and flew the Hump in the CBI (China, Burma, India) theatre of operations. Post-war it served on the Berlin Airlift with both the US Air Force and the US Navy, and played a very prominent role in the United Nations conflict in Korea.

Over 200 examples of the Douglas Skymaster transport went into service with the US Navy and the US Marine Corps units during World War II, many remaining in service twenty years later. Fortunately all five US Navy variants were standardised with the USAF variants and were designated Douglas R5D and served with distinction with squadrons of the Naval Air Transport Service (NATS) between 1941-48, later assigned to the huge Military Air Transport Service (MATS) and to the Fleet Logistics Air Wing and US Marine Corps units.

The Douglas R5D-1 was the equivalent to the C-54A with R-2000-7 engines, some fifty-eight being acquired by the US Navy. Thirty R5D-2s, equivalent to the C-54, had revised fuel system, while the ninety-eight R5D-3 transports had R-2000-11 engines, being similar to the C-54D. Twenty R5D-4s were equivalent to the C-54E, and a minor engine change, to R-2000-9, plus a revised interior, identified the R5D-5. The R5D-6 designation indicated a mythical equivalent of the cancelled C-54J with airline-type interior. There were a number of odd-balls — designations like R5D-4R, a basic cargo carrier modified as a personnel carrier. Douglas R5D-5Z indicated a VIP equipped transport, one of which was initially used by the US Coast Guard.

On 6 July 1962 under a Department of Defense Directive the R5D designations allocated to the US Navy Skymasters were standardised with those of the many US Air Force variants. This affected some of the Skymasters in service with the US Coast

Seen in flight and carrying the minimum of US Coast Guard identity is Douglas R5D-3 BuNo 72467 on 30 January 1951. Some fifteen R5D-3/4 Skymasters served the USCG in a variety of roles including International Ice Patrol, LORAN station calibration, SAR etc. The type served in Newfoundland, Hawaii and air stations in the US. *(USCG)*

Guard, resulting in a mixture of designations being recorded on the aircraft, presenting some real puzzlers to the student of aviation history. The majority of USCG aircraft were regular C-54D and C-54G models, with the odd EC-54U indicating electronic testing equipment and RC-54V when cameras were fitted. It will be noticed by the serial listing of the USCG transports that dual serials — US Navy BuNos and USAF — were issued for some aircraft.

The aircraft used by the Coast Guard were employed on a wide variety of missions including Search & Rescue, service with the International Ice Patrol, photo mapping, electronic tests, and were based at the USCG air stations and detachments located at Elizabeth City, North Carolina; Barbers Point, Hawaii; San Francisco, California; and Argentia, Newfoundland. Since 1914 the USCG has been tasked with operating the International Ice Patrol in the North Atlantic and has proved to be an outstanding example of international collaboration. Each year, from March to September, Coast Guard aircraft patrol a 33,000 square mile area off the Grand Banks, a rugged section of the North Atlantic, crossed by the busiest shipping lanes in the world, which is also the natural route for thousands of icebergs that break off the western coast of Greenland in spring and are carried south on the Labrador Current. Three Coast Guard transports are known to have carried 'ARGENTIA' on the tail and operating the ice patrol, these including 2468 and 5614.

A retired USCG Captain and aircraft commander once told the author that on accepting a R5D transport from the US Navy it still had remnants of coal dust indicating it had served on the Berlin Airlift. One transport, 2623, had served with the USAF as JC-54D on the Atlantic Missile Range. Three aircraft, 2451, 5532 and 5583, were returned for storage at the Davis-Monthan boneyard in Arizona. Douglas R5D-5Z was seen by the author during 1954 at the US Naval Air Facility, Blackbushe in the United Kingdom. This later became an electronics aircraft based at Elizabeth City with the air station name on the fin. Others based here included 2486 and 5614.

Three transports were recorded at the San Francisco air station, these including 5587 and 2486 on 27 December 1958 and 5490 on 27 September 1959, the latter in new high-visibility markings. On 18 August 1964 two aircraft 5540 and 5576 were sold at Elizabeth City to a civil operator, while the last mention in the USCG aircraft inventory of a C-54 Skymaster is dated 1 June 1965, recording 9147 with Gary Aircraft, Victoria, Texas, presumably awaiting disposal.

TECHNICAL DATA

Manufacturer:	Douglas	Engine:	Pratt & Whitney R-2000-7
Other designations:	C-54	Take-off power:	4 × 1290 hp
Span:	117 ft 6 in	Designation:	R5D-3/4/5 Skymaster
Height:	27 ft 6 in	Type:	Transport
Fuel:	3580 galls	Length:	93 ft 10 in
Empty weight:	37,000 lb	Wing area:	1460 sq ft
Crew:	4	Gross weight:	62,000 lb
Top speed:	265 mph	Range:	3900 st miles
Cruise:	210 mph	Service ceiling:	22,000 ft
Sea-level climb:	148 ft per min to 10,000 ft		

DOUGLAS R5D-3/4 SKYMASTER

USCG No.	Tail identity & designation	Model	c/n	USAF/USN Serial
9147	R5D-5Z 49147	EC-54U-20-DC	27373	44-9147
5614	R5D-3 (fuselage)	C-54G-10-DO	36067	45-614
2451	R5D-3 (fuselage)	C-54D-1-DC	10556	42-72451
2486	R5D-3 (fuselage)	RC-54V-1-DC	10591	42-72486
2623	R5D-3 72623	C-54D-DC	10728	42-72623
2467	R5D-3 72467	C-54D-DC	10572	42-72467
5490	R5D-4 (fuselage)	C-54G-DO	35943	45-490
7227		C-54D-1-DC	10556	42-72451
5532		C-54G-5-DO	35985	45-532
5540		C-54G-5-DO	35993	45-540
5576		C-54G-10-DO	36029	45-576
5583		EC-54U-10-DO	36036	45-583
5587		C-54G-109-DO	36040	45-587
50877	R5D-3	C-54G	10563	42-72458
90410	R5D-4R	C-54G	27366	44-9140

Notes: DO = Douglas-built at Santa Monica
DC = Douglas-built at Chicago

FAIRCHILD J2K-1/2

The Fairchild Model 22 XR2K-1 of 1936 was a two-seat commercial parasol monoplane powered by a 145hp Warner radial engine. It was procured by the US Navy as a research vehicle for the National Advisory Committee for Aeronautics (NACA) and never operated as a naval service aircraft. The 'K' in the designation identified Kreider-Reisner, a company which Fairchild had absorbed in 1929, but managed to still maintain its identity on production aircraft. The XRK-1 designation had already been assigned to Kreider.

Other Kreider-Reisner designs purchased by the US Navy were JK-1, J2K-1 and J2K-2, which were commercial Fairchild 24-R. Two J2K-1s and two J2K-2s were purchased by the Coast Guard during 1936. These were three-seat models powered by inverted air-cooled Ranger 145hp engines.

The two Fairchild J2K-1 aircraft were assigned to the USCG air station located at St Petersburg, Florida, while the two J2K-2s were assigned to the air station at Charleston, South Carolina. All were unfortunately lost in accidents.

Four Fairchild J2K-1/2 high-wing monoplanes were acquired by the USCG during May 1937, but by the early years of World War II all had been lost in accidents. Depicted is J2K-1 'V160' which was purchased and restored by Carl Swickley and registered N81234. *(Peter M Bowers)*

TECHNICAL DATA

Manufacturer:	Fairchild	Prop diameter:	7 ft 6 in
Other designations:	F-24H	Designation:	J2K-1/2
Contract No.:	Tcg. 26669	Type:	Observation
Span:	36 ft 4 in	Unit cost:	$6466 & $7123
Height:	8 ft	Length:	24 ft 10 in
Fuel:	240 galls	Wing area:	173 sq ft
Empty weight:	1560 lb	Oil:	6 galls
Crew:	4	Gross weight:	2550 lb
Top speed:	138 mph	Stall speed:	47 mph
Cruise:	127 mph	Range:	560 miles
Sea-level climb:	560 ft per min	Service ceiling:	16.500 ft
Engine:	Ranger 6-410	Two-blade prop blades:	Sensevich-wood
Take-off power:	1 × 165 hp		

SERIAL INFORMATION

USCG No.	Commissioned	Decommissioned	
J2K-1 V160	Mar 1937	Aug 1940	Crashed
V161	Mar 1937		
J2K-2 V162	May 1937	May 1941	Crashed
V163	May 1937	Aug 1939	Crashed

FAIRCHILD C-123B PROVIDER

A medium assault cargo aircraft initially built for the US Air Force by Chase in 1953 at Willow Run, Texas, the Fairchild C-123B Provider was used by the US Coast Guard for logistic support of the far-flung empire of the many USCG operated LORAN (long-range navigation) stations. Eight of these twin-engined high-wing transports were operated, being based at remote Coast Guard air detachments and air stations located at Guam in the Marianas; Honolulu, Hawaii; Kodiak, Alaska; San Juan, Puerto Rico; and Naples, Italy. Powered by two Pratt & Whitney R2800 engines, these transports could carry a useful load of cargo to the isolated outposts.

The C-123 entered US Air Force service during July 1955, after a long, interesting and rather complicated period of development, the basis of the design originally commencing with the XG-20 cargo glider produced by Chase Aircraft in 1949. Of all-metal construction, it was originally designed for adaption as a powered assault transport. The first prototype flew powered with R2800-83 engines fitted in wing nacelles. This became redesignated XC-123 Avitruc. A second prototype fitted with four J47 turbojets in paired pods under the wing became the XC-123A, first flown on 21 April 1951. During 1952 the US Air Force placed contracts with Chase for five pre-production C-123B transports. The following year, in 1953, the Kaiser Frazer Corporation acquired a majority interest in Chase, resulting in a production contract for 300 aircraft being awarded.

After Chase had built and flown its five C-123Bs from Willow Run, Texas, in 1953, difficulties encountered by Kaiser Frazer led to the cancellation of their contract on 24 June 1953 in favour of a new contract placed with the Fairchild Engine & Airplane Corporation of Hagerstown, Maryland, later the same year. This company took over the responsibility for the continued flight development of the Chase-built C-123Bs, and introduced on these a large dorsal fin which became standard on production Providers.

The first Fairchild-built C-123 first flew on 1 September 1954, production totalling 302 on US Air Force contracts, including one static test airframe plus twenty-four for Mutual Aid Procurement (MAP) to Venezuela and Saudi Arabia. A single VC-123C command transport version planned by Kaiser Frazer was not proceeded with.

A look at the US Department of Defense aircraft designations in the 'C' for Cargo category reveals that the C-123 designation was also carried by two similar versions developed by Stroukoff Aviation, a company formed by the original designer of the aircraft when he was employed by Chase Aircraft. These were the YC-123D, which had boundary layer control by means of suction slots in the upper wing surfaces; and the YC-123E with 'Pantobase' undercarriage for operations from sand, snow, ice, water and land. Stroukoff had modified the first C-123B to the YC-134, intended to test a boundary layer control system. This prototype had Wright R-3350-89A turbo-compound engines fitted with four-bladed propellers, an enlarged fuselage, and a much modified undercarriage. Small end-plate fins and rudders replaced the dorsal fin. When fitted with 'Pantobase' hydro-skis it was designated YC-134A.

Many C-123 Providers served in Vietnam with the US Air Force where they were used in a wide variety of roles. These included defoiliation, night interdiction and reconnaissance, while a single VC-123K was the personal transport of General Westmoreland. Air America used the C-123 Provider on a wide variety of services — some clandestine — on behalf of the US government in South East Asia for almost two decades.

Early in 1958 the first of eight Fairchild C-123B Providers was acquired by the US Coast Guard, the remaining seven being delivered during early 1961, in fact these seven were the only

Fairchild C-123B Provider '4505' seen in flight over the Pacific from its home base in Honolulu, Hawaii. One of eight operated by the USCG for logistic support of its PORAN stations worldwide and based with USCG air detachments. The last Provider was retired in June 1952 by the Coast Guard. *(USCG)*

aircraft acquired by the USCG in that year. The final C-123B was retired from Coast Guard service in June 1972. The USCG aircraft inventory dated 1 February 1965 gives the location of the transports: Naples, Italy, operated 4540 and 4357; San Juan, Puerto Rico, had 4705, this being changed for 4358 by June 1965; Elizabeth City, North Carolina, used 4705 for support logistics, while Kodiak, Alaska, had 4529 and 4668 and Guam in the Marianas operated 4505 and 4541. By July 1971, San Juan had given up its C-123Bs while Miami, Florida, operated two, 4505 and 4705.

It was normal for the USCG to have six C-123Bs assigned retaining two, possibly under overhaul. In Europe alone the USCG is still responsible for some six LORAN stations, all manned by a nucleus of USCG personnel and controlled from the USCG European Activities office in London. The location of these stations ranges from Iceland to Turkey.

TECHNICAL DATA

Manufacturer:	Fairchild	Take-off power:	2 × 2300 hp
Span:	110 ft	Designation:	C-123B Provider
Height:	34 ft 1 in	Type:	Supply transport
Empty weight:	29,900 lb	Length:	75 ft 9 in
Crew:	3	Wing area:	1223 sq ft
Top speed:	245 mph	Gross weight:	60,000 lb
Cruise:	205 mph	Range:	1470 st miles
Sea-level climb:	1150 ft per min	Service ceiling:	29,000 ft
Engine:	Pratt & Whitney R-2800-99W		

Miscellaneous information:

4668	C-123B-9-FA	54-668
4705	C-123B-10-FA	54-705
4505	C-123B-12-FA	55-4505
4529	C-123B-13-FA	55-4529
4540	C-123B-14-FA	55-4540
4541	C-123B-14-FA	55-4541
4357	C-123B-17-FA	56-4357
4358	C-123B-17-FA	56-4358

GENERAL AVIATION PJ-1/2 FLYING LIFE BOAT

Known for many years by US Coast Guard aviators as FLBs — Flying Life Boats — the five flying-boats manufactured specifically for the service marked the swan song of the American Fokker organisation. The sleek twin-pusher flying-boats were designed by the Fokker Aircraft Corporation of America as Model AF-15 in a Coast Guard design competition involving eight companies for a patrol and open-sea rescue boat. By the time that Fokker with its FLB was declared the winner, the company had become a division of the General Aviation Corporation with manufacturing facilities transferred to Dundalk, Maryland. These flying-boats were delivered during 1932, and for many years were simply known as the 'Flying Life Boats'. This abbreviated as FLB appeared along with the USCG serial number on the tail of each, serving as a model designation. The first aircraft was erroneously marked FLB-8 when it was commissioned in January 1932, but this was soon changed to the correct FLB-51. Later, when General Aviation became North American, the PJ-1 model designation was adopted, this being Fokker/General Aviation F2B re-designated.

The first flying-boat, FLB-51, named *Antares* was completely modified in 1933, being sent to the Naval Aircraft factory and converted to a tractor design. The engine nacelles and cowlings were changed, and the engine mounts, pilot's cockpit and other sections were modified. This changed the model designation for this one aircraft to PJ-2. Like the Douglas RD Dolphins the fleet of five were named after the stars, the name appearing on each side of the nose. Often this was the only accurate means of identification of the individual aircraft. The FLBs had a separate radio and a radio direction finding instrument included in their layout.

OPERATIONS

Late in 1932, the five Flying Life Boats became operational, and were very substantial aircraft for their day. They had a gross weight of just over 11,000lb and a range of 1000 nautical miles at 120mph. Rescue exploits by these aircraft were legendary.

A passenger on board the US Army transport *Republic* en-route from Panama to New York became critically ill. An emergency operation beyond the transport's facilities was necessary. Help was sought. The PJ-1 *Arcturus* was lowered down the ramp at the USCG air station at Miami, taxied for take-off into Biscayne Bay and flew three hours in darkness and a storm. Using radio bearings, the searchlight beams from the *Republic* were located and the flying-boat circled and landed off the ship's bow. The sick patient and his wife were transferred by one of the ship's lifeboats. *Republic* indicated the wind direction for take-off by its searchlights and *Arcturus* lifted off out of the

Known for many years, and in fact designated Flying Life Boat (FLB), the General Aviation PJ-1/2 flying-boats operated by the USCG were delivered during 1932, some serving into the early war years. Depicted in flight is PJ-1 'FLB-54' *Acamar* which was based at Miami in 1933. *(USCG)*

rough sea. This difficult mission took seven hours — and a life was saved.

During another rescue incident involving PJ-1, Lieutenant Commander Carl von Paulsen set the *Arcturus* FLB-55 down in a heavy sea during January 1933, off Cape Canaveral, Florida, and rescued a boy adrift in a skiff. The flying-boat sustained so much damage during the open water landing that it was unable to take off. This was the fate on a number of ocean rescues which had to be attempted when no other rescue craft could be directed to the scene by the aircraft. Ultimately, the *Arcturus* washed onto the beach and all, including the boy, were saved.

TECHNICAL DATA

Manufacturer:	General Aviation	Take-off power:	2 × 420 hp (PJ-2 500 hp)
Other designations:	FLB	Designation:	PJ-1/2 Flying Life Boat
Contract No.:	Tcg 12154	Type:	Flying Life Boat
Span:	74 ft 2 in	Unit cost:	$73,343
Height:	15 ft 6 in	Length:	55 ft
Fuel:	440 galls	Wing area:	754 sq ft
Empty weight:	7000 lb	Gross weight:	11,200 lb
Crew:	4	Passengers:	3
Top speed:	120 mph (PJ-2 135 mph)	Range:	1000 nm
Engine:	Pratt & Whitney Wasp C-1 R-1340	Prop blades:	Hamilton Standard Fixed-pitch

Miscellaneous information:
PJ-2 Pratt & Whitney T1D1 Hornet R-1690.

SERIAL INFORMATION

USCG No	Commissioned	Decommissioned		
FLB-51 251 V116	June 1932		PJ-2	*Antares*
FLB-52 252 V112	Aug 1932	May 1940	PJ-1	*Altair*
FLB-53 253 V113	Sep 1932	Oct 1940	PJ-1	*Acrux*
FLB-54 254 V114	Sep 1932	Aug 1937	PJ-1	*Acamar*
FLB-55 255 V115	Nov 1932	Aug 1941	PJ-1	*Arcturus*

Miscellaneous information:

Antares	Radio call-sign	'NUMRL'	Coast Guard	'24 L'	Cape May	1933
Altair		'NUMRM'		'24 M'	Cape May	1933
Acrux		'NUMRN'		'24 N'	Cape May	1933
Acamar		'NUMRO'		'24 O'	Miami	1933
Arcturus		'NUMRP'		'24 P'	Miami	1933

GRUMMAN JF-2 J2F-5/6 DUCK

During 1931 the US Navy re-assigned the letter 'J' in their aircraft designation system to identify a new category of aircraft intended as a general utility type for assignment to utility flights aboard aircraft carriers. The Grumman G-7 single-engined amphibian had the distinction of becoming the first type to be designed to meet the US Navy requirements for a general utility aircraft. Preliminary work on the design commenced in the summer of 1931, and a prototype was ordered as the XJF-1 in 1932, the first flight being made on 24 April 1933.

Some sixteen months went by between the time Grumman first discussed its proposed utility landplane or amphibian and the ordering of the XJF-1 prototype, as the US Navy did not have funds immediately available. During this long period, requirements were refined until the need to stress the aircraft for catapult launch from warships was eliminated from the specification.

The Grumman Duck utility amphibian was first ordered into quantity production during early 1934 when the US Navy Department awarded contract No 32111 for twenty-seven JF-1s. Three other JF variants were then built for the US Navy, the US Marine Corps, and the US Coast Guard. An enlarged and more versatile variant of the Duck, the Design G-15 J2F-1 was first ordered in 1935, with subsequent orders resulting in the production of four additional models for the US military services including the Coast Guard. To enable Grumman to concentrate on the production of fighters and torpedo-bombers at Farmingdale, New York, manufacture of the Duck was transferred in 1942 to the Columbia Aircraft Corporation in Valley Stream, Long Island, New York.

The later Grumman J2F-5 and J2F-6 were similar in dimensions and performance to the earlier JF-2. Five were acquired by the Coast Guard between January 1942 and October 1945. By mid-1946 the USCG commenced returning the type to the US Navy, to the US Air Force and the huge War Assetts Administration. One Duck was lost during a storm in October 1946.

In addition to their normal duties, Coast Guard aviators were setting up various records with their aircraft. On 20 December 1934 Commander Elmer F Stone set up a record of 191.734mph in a Grumman JF-2 V167 at Buckroe Beach, Virginia, when he flew a set three-kilometre course. On 25 June 1935 Lieutenant Burke, in another Grumman JF-2, set a further record of 173.945mph over a 100-kilometre course carrying a 500kg load.

OPERATIONS

There were losses to both men and machines. On 29 September 1940 the Coast Guard lost Lieutenant T G Miller and Seaman 2nd Class T B Redman when their Grumman JF-2 V145 crashed in the ocean near St Petersburg, Florida, while on a local night training flight from the air station.

Preliminary flight training and indoctrination was given to officers and enlisted men at the USCG air station located at Charleston, South Carolina, during early 1941 prior to their selection for flight training at NAS Pensacola, Florida, with the US Navy, which had allocated the Coast Guard a quota of ten men to be trained between January and June 1941. In order to speed up this flight training programme two Grumman JF-2s V135 and V141 were transferred to the US Navy in exchange for three Naval Aircraft Factory N3N-3 trainers.

Seen over the mountainous terrain of Washington State, is Grumman JF-2 Duck 'V148' which was delivered to Port Angeles Air Station from the factory on 21 November 1935. The type operated from USCG cutters, and during 1939 'V148' was detached to the Bering Sea Patrol.
(Gordon S Williams)

Ten Grumman Duck utility amphibians were acquired by the Coast Guard from the US Navy. Described as a very rugged aircraft, apparently the J2F had a mind of its own when landing on a conventional runway with a crosswind. The last one was retired from the Coast Guard during 1948.

On 1 November 1941 a total of nine Coast Guard air stations and fifty-six aircraft were transferred to the US Navy jurisdiction.

Four aircraft were employed on special assignments, including JF-2 V135 based on the USCG cutter *Taney*, and JF-2 V141 on detachment at Charleston, South Carolina. The USCG aircraft assignment listing dated 31 December 1941 indicated that Grumman JF-2 Ducks V137 and V143 were assigned to Biloxi air station, Mississippi, while San Diego, California, had JF-2 V139; San Francisco JF-2 V140 and Port Angeles in Washington state operated JF-2 V148.

A similar aircraft assignment listing dated 28 February 1943, indicated that Elizabeth City air station, North Carolina, had on strength Grumman J2F-5 Duck BuNo 00751, while Port Angeles, Washington, had J2F-5 BuNo 00796 and J2F-4 BuNo 1667 on their inventory.

In its variety of versions, the Grumman Duck served throughout World War II providing the manufacturer with more than a solid basis for development of its later amphibian for Coast Guard use. After World War II in 1946/47 during Operation High Jump the great US Navy armada that ventured as a task force to Antarctica included ships and aircraft from the US Coast Guard. Grumman J2F-6 Duck seaplanes were detached to the USCG icebreaker *Northwind*, being used for making reconnaissance flights over the South Pole regions.

TECHNICAL DATA

Manufacturer:	Grumman	Prop diameter:	9 ft
Contract No.:	Navy 33862 40569	Designation:	JF-2 J2F-5/6 Duck
Span:	39 ft	Type:	Utility amphibian
Height:	12 ft 4 in	Unit cost:	$45,000
Fuel:	150 galls	Length:	34 ft
Empty weight:	4114 lb	Wing area:	409 sq ft
Crew:	2	Oil:	11 galls
Top speed:	176 mph	Gross weight:	5800 lb
Cruise:	155 mph at 7000 ft	Stall speed:	67 mph
Sea-level climb:	1500 ft per min	Range:	759 miles at 7000 ft
Engine:	Wright Cyclone R-1820-102	Service ceiling:	18,500 ft
Take-off power:	1 × 775 hp	Three bladed prop blades:	Hamilton Standard

GRUMMAN JF-2 DUCK

USCG No.	c/n	Delivered	Base	Ferry pilot
161 V135	188	27 Oct 1934	Gloucester	Cdr von Paulsen
161 V136	189	2 Nov 1934	Cape May	Lt R L Burke
163 V137	190	9 Nov 1934	Cape May	Lt R L Burke
164 V138	191	9 Nov 1934	Cape May	Lt Wm Schissler
165 V139	192	23 Nov 1934	Cape May	Lt W Sinton
166 V140	193	23 Nov 1934	Cape May	Lt Wm Schissler
167 V141	194	23 Nov 1934	Cape May	Lt E E Fahey
168 V142	195	3 Dec 1934	Cape May	Lt W Sinton
169 V143	196	2 Jan 1935	Cape May	Lt E E Fahey
170 V144	263	29 Oct 1935	Port Angeles	Lt C F Edge
171 V145	264	4 Nov 1935	St Petersburg	Lt W A Burton
172 V146	265	8 Nov 1935	Salem	Lt W L Foley
173 V147	266	13 Nov 1935	Miami	Lt C B Olsen
174 V148	267	21 Nov 1935	Port Angeles	Lt C F Edge
175	268	26 Nov 1935	Quantico, VA	Lt H D Palmer

The last entry above is historically very important as it confirms that the Coast Guard traded to the US Marine Corps at Quantico, Virginia, Grumman JF-2 Duck 175 for the Lockheed R30-1, later V151.

In January 1941 four USCG JF-2 Ducks were traded to the US Navy — V135 to BuNo 0266; V141 to BuNo 00371; V144 to BuNo 00372 and V146 to BuNo 01647 — in exchange for four Naval Aircraft Factory N3N-3s. V136 crashed in January 1940; V142 crashed in July 1936; V147 crashed in October 1939*. V135 served on USCG cutter *Taney* in Honolulu, Hawaii; V144 served on USCG cutter *Spencer* in Cordova, Alaska. V148 served on the Bering Sea Patrol during 1939.

BuNo 00751 based at Elizabeth City, North Carolina 28 Feb 1943 J2F-5
BuNo 1667 based at Port Angeles, Washington 28 Feb 1943 J2F-4
BuNo 00796 based at Port Angeles, Washington 28 Feb 1943 J2F-5
BuNo 33585 attached to USCG icebreaker *Northwind* Operation High Jump 1946.

* V145 crashed at St Petersburg on 29 September 1940 on a night training flight. Crew of two killed.

GRUMMAN JRF-2/3/5G GOOSE

The Grumman G-21 Goose became, on 29 May 1937, the first Grumman monoplane to fly. Early in 1930 the company had proposed a twin-engined, parasol monoplane flying-boat for the US Coast Guard under the company design G-3. It was not accepted by the Coast Guard and remained on the drawing board.

Work was initiated during 1936 to meet the needs of wealthy aircraft owner businessmen. With hydrodynamicist Ralston Stalb designing a two-step hull, the design evolved into a clean monoplane powered by a pair of Pratt & Whitney Wasp Jr nine-cylinder radials which were mounted on the wing leading edge, forward of the cabin, and fitted with collector rings exiting above the wings to reduce cabin noise. Accommodation was provided for two pilots situated in a cockpit forward of the wing and for four to six passengers in a roomy cabin located beneath the high-mounted wing. Entrance was via a door on the port side just aft of the cabin and baggage was stowed in a bow compartment and after of the cabin.

Other distinguishing features of the Goose included its hand-cranked retractable undercarriage which was used for land operation or could be extended in the water during beaching operations. The main wheels retracted upward into the side of the fuselage, as in earlier Grumman aircraft, and a steerable tailwheel retracted aft into the hull behind the second step. For land operations, the wing floats could be removed, thus reducing drag and increasing performance. During operations from snow covered areas, the main undercarriage and tail wheel could be replaced with skis.

The ease of handling, good stability, and satisfactory performance demonstrated during flight trials soon made the Grumman Goose a very popular aircraft with both civil and military customers. It proved to have a very strong airframe, thus endowing many of the 345 aircraft built by Grumman between May 1937 and October 1945 with a long service life.

Seven G-39 design amphibians were built to meet Coast Guard requirements. Designated JRF-2 they were powered by two 450hp Wasp Jr SB-2 radial engines. The passenger seats were replaced when required by stretchers. The aircraft carried USCG serials V174/175/176. Three JRF-3 Goose amphibians bearing USCG serials V190/191/192 were delivered during November 1940, and differed from the JRF-2 in being fitted with an autopilot and with de-icing boots on the leading edge of their wings and tail surfaces.

Beginning in 1941, Grumman commenced delivery of the G-38 design JRF-5, the principal production version of the Goose. This had uprated engines, cameras for air survey work and other refinements. Production totalled 184, including two dozen supplied to the Coast Guard as JRF-5G. The engines were two 450hp R-985-AN-6 radials manufactured by Pratt & Whitney.

By the end of 1941, the US Navy had thirty-five Goose amphibians which included the XJ3F-1, four JRF-1 transports, five JRF-1A utility aircraft, ten JRF-4s and fifteen JRF-5s which could be used in the anti-submarine role. The Coast Guard had seven JRF-2s and three JRF-3s. During World War II, when the Coast Guard and the US Navy operated side by side and exchanged aircraft, 169 JRF-5s were added to the inventory plus four G-21A Goose taken from civil operators. The US Navy and Coast Guard combined force used these versatile amphibians for utility duties, light transport, and on coastal anti-submarine patrols. The US Navy disposed of its JRFs shortly after the end of hostilities, while the Coast Guard retained some of its JRF-5Gs until the mid-1950s.

OPERATIONS

On 27 January 1942 a dual role of bombing the enemy from the air and bringing aid to survivors was once more played by Coast Guard Grumman JRF-2 Goose V175, piloted by Lieutenant Commander R L Burke from the Elizabeth City air station in North Carolina. The JRF received a distress call from the 7096-ton US tanker *Frances E. Powell* indicating that she was being overtaken by a submarine eight miles off Currituck Light south of Virginia Beach. The tanker was sunk soon afterward. A USCG Grumman J2F-5 Duck first sighted the submarine and dropped two depth charges within a hundred feet of the now submerged marauder. Then Commander Burke in V175 dropped a grapnel with a hundred feet of line and two life jackets to buoy the spot so that destroyers could later depth-charge the area. It seemed likely that the enemy had been damaged because Burke later saw and photographed what appeared to be a distress buoy from a submarine. All but four of the tanker's crew of thirty-two were eventually saved by surface craft summoned to the scene by the two Coast Guard aircraft.

The US Coast Guard aircraft inventory dated 17 December 1941 gives the location of USCG JRF Goose amphibians. JRF-2 V185 was based at Salem air station, Massachusetts; Floyd Bennett Field, Brooklyn, New York, operated JRF-2s V190 and V191; Elizabeth City, North Carolina, had JRF-2 V175 and V186; Biloxi, Mississippi, operated a single JRF-2 V184; Port Angeles, Washington, had JRF-2 V176, while the Air Patrol Detachment located at Traverse City, Michigan, had JRF-3 V192 on charge; the Senior Coast Guard officer of the 14th Naval District had on charge JRF-2 V187, and the Senior Coast Guard Officer at the 7th Naval District had JRF-2 V174.

A similar USCG aircraft assignment listing dated 28 February 1943 gave the following location for JRS Goose amphibians: Salem JRF-2 V185, JRF-3 V192; Brooklyn JRF-3 V191; Elizabeth City JRF -2 V186; Biloxi JRF-2 V174 and V184; Port Angeles JRF-2 V175 and V176.

A Coast Guard Grumman JRF- Goose in World War II camouflage seen on patrol over the desolate and rugged terrain of Alaska on 27 July 1944. The type was a very rugged aircraft and could carry depth charges and bombs on racks under the wings. The Coast Guard operated the Goose into the 1950s. *(USCG via William T Larkins)*

As part of the US Navy seaplane research programme, the Bureau of Weapons funded the testing of a Gruenberg supercavitating hydrofoil system. A single hydrofoil was fixed beneath a modified Grumman JRF-5G BuNo 37782/US Coast Guard 7782. Small hydro-skis were attached to slanted attachments projecting beneath the bow, and the two-blade propellers were replaced by three-blade units. The practicality of the concept was successfully demonstrated during the summer of 1962, but by then both the US Navy and the US Coast Guard were fast losing interest in flying-boats and no further development was funded.

The Coast Guard Goose amphibians were utilised in every kind of duty imaginable before, during and after World War II. They carried depth charges, bombs, passengers, mail, picked up ditched aircrew, stranded mariners and even USO troupes. They were a very rugged utility aircraft.

TECHNICAL DATA

Manufacturer:	Grumman	Prop diameter:	8 ft 6 in
Other designations:	Army OA-13B	Designation:	JRF-2/3 Goose
Contract No.:	Tcg 29648	Type:	Utility transport amphibian
Span:	49 ft	Unit cost:	$79,526
Height:	12 ft	Length:	38 ft 4 in
Fuel:	220 galls	Wing area:	376 sq ft
Empty weight:	5530 lb	Oil:	15 galls
Crew:	6	Gross weight:	7955 lb
Top speed:	200 mph at 5000 ft	Stall speed:	65 mph
Cruise:	150 mph	Range:	1000 miles
Sea-level climb:	1300 ft per min	Service ceiling:	22,000 ft
Engine:	Pratt & Whitney Wasp Jr R-985-AN6	Two-blade prop blades:	Hamilton Standard, two position
Take-off power:	2 × 400 hp		

TECHNICAL DATA

Manufacturer:	Grumman G-21	Prop diameter:	8 ft 6 in
Other designations:	Army OA-13B	Designation:	JRF-5G Goose
Span:	49 ft	Type:	Utility transport amphibian
Height:	15 ft	Unit cost:	$70,950
Fuel:	220 galls	Length:	38 ft 4 in
Empty weight:	6415 lb	Wing area:	375 sq ft
Crew:	2	Oil:	15 galls
Top speed:	180 mph	Gross weight:	8700 lb
Cruise:	150 mph at 1500 ft	Stall speed:	65 mph
Sea-level climb:	1300 ft per min	Range:	800 miles at 1500 ft
Engine:	Pratt & Whitney Wasp Jr R-985-AN6	Service ceiling:	22,000 ft
Take-off power:	2 × 450 hp	Two-blade prop blades:	Hamilton Standard, 2D30-235

SERIAL INFORMATION

	USCG		Commissioned	Decommissioned	
JRF-2	V174	c/n 1063	Jul 1939	Jan 1947	Sold as surplus
	V175	1064 To US Navy as BuNo 0266.			
	V176	1065	Oct 1939	Apr 1943	Crashed
	V184	1076	Feb 1940		
	V185	1077	Feb 1940	Nov 1946	Sold
	V186	1078	Mar 1940	Aug 1946	
	V187	1079			
JRF-3	V190	1085	Apr 1940	Aug 1946	Crashed
	V191	1086	Dec 1940	Nov 1946	
	V192	1087	Nov 1940	Oct 1948	
JRF-5G	V224	BuNo 34079	Aug 1943	May 1954	Returned to US Navy
	V225	34082			
	V226	34087	Oct 1943	Oct 1946	Crashed
	V227	34090	Oct 1943	Aug 1953	Stored
	V228	37772	Nov 1943	Jun 1954	Returned to US Navy
	V229	37773	Nov 1943	May 1954	Returned to US Navy

BuNo 84792 USCG 4792 San Francisco air station September 1951
37794 USCG 794 Anchorage, Alaska 1971
87733 USCG 7733 Salem, Massachusetts 31 July 1952

GRUMMAN J4F WIDGEON

The Grumman XG-44 prototype was first flown by Bud Gillies and Roy Grumman at Bethpage, New York, on 28 June 1940. Flight trials proceeded slowly but satisfactorily, the only modification required before the issue of an Approved Type Certificate on 5 April 1941 consisted of adding elevator horn balances.

After receiving its certificate, the G-44, named Widgeon, entered a programme with forty-one aircraft being delivered during 1941 to civil customers and the Portuguese naval air arm. Production then switched to J4F-1s and J4F-2 for both the Coast Guard and the US Navy.

Originally ordered as a utility aircraft, twenty-five J4F-1s were built for the Coast Guard in two batches — V197 to V204 delivered in 1941, followed by V205 to V221, delivered during the following year. They were powered by 200hp Ranger L440·5 engines and initially differed from the civil aircraft in having an upper fuselage hatch just behind the wing to enable easy loading and unloading stretchers. Later, a rack was added beneath the starboard wing to carry either a 200lb depth charge, or a raft and rescue gear.

The Grumman J4F- Widgeon utility amphibian served during World War II with the USCG, twenty-five being delivered during 1941/42. Later they were equipped to carry depth charges or packs of rescue gear under the wing. Depicted is J4F-1 'V203' from Port Angeles Air Station, Washington. *(USCG)*

OPERATIONS

In Coast Guard service the Widgeons were mostly used for search and rescue in coastal waters and as utility transports. During the early years of World War II they were also used on anti-submarine patrols. On 1 August 1942 a Grumman J4F-1 from Coast Guard Patrol Squadron 212, based at Houma, Louisiana, and flown by Chief Aviation Pilot Henry C White, with his crewman Radioman First Class George Henderson Boggs Jr were patrolling an area near a sunken ship off the Passes of the Mississippi in the Gulf of Mexico. White sighted a submarine achored on the surface. He felt the most effective attack would be from astern, but while he was circling to make the attack, the submarine began to submerge. White commenced his attack immediately from abeam when he was still at an altitude of 1500ft and half-a-mile from the U-boat. The submarine was submerging quickly, so White put his Widgeon amphibian in a 50 degree dive, releasing his bomb from an altitude of 250 feet. Although Boggs witnessed the bomb strike the water and explode in what appeared to be a direct hit, no debris was spotted. Patches of oil coming to the surface, though, indicated that heavy damage had been sustained. It was subsequently learned that *U-166* had been completely destroyed — the only enemy submarine sank by a Coast Guard aircraft during World War II. White was subsequently awarded a Distinguished Flying Cross.

The Coast Guard aircraft assignment document dated 28 February 1943 gives the following assignment information for Widgeon aircraft. The air station at Salem, Massachusetts, operated two J4F-1s V197 and V218; Brooklyn, New York, operated no less than four J4F-1s V198, V205, V210 and V215; St Petersburg, Florida, operated four J4F-1s V200, V209, V213 and V216; Biloxi, Mississippi, operated five J4F-1s V202, V211, V212, V214 and V217; San Diego, California, had three J4F-1s V204, V219 and V220; San Francisco had a single J4F-1 V201 and Port Angles, Washington, had four J4F-1s V203, V206, V207 and V221. Elizabeth City, North Carolina, operated a single J4F-1 V208.

TECHNICAL DATA

Manufacturer:	Grumman G-44	Prop diameter:	6 ft 10 in
Contract No.:	Tcg 33459	Designation:	J4F-1 Widgeon
Span:	40 ft	Type:	Light amphibian
Height:	8 ft 10 in	Unit cost:	$75,526
Fuel:	108 galls	Length:	31 ft 1 in
Empty weight:	3225 lb	Wing area:	245 sq ft
Crew:	4	Oil:	7 galls
Top speed:	153 mph	Gross weight:	4525 lb
Cruise:	138 mph	Stall speed:	61 mph
Sea-level climb:	870 ft per min	Range:	780 nm
Engine:	Ranger L-440-5	Service ceiling:	14,600 ft
Take-off power:	2 × 200 hp	Two-blade prop blades:	Sensenich 32-RS-72

GRUMMAN J4F-1 WIDGEON

USCG Nos.	c/n	Commissioned	Decommissioned	Notes
V197	1222	Jul 1941	Jan 1944	
V198	1223	Jul 1941	Nov 1946	
V199	1224	Jul 1941	Dec 1941	Crashed
V200	1225	Aug 1941		
V201	1226	Aug 1941	Aug 1947	Surplus. N69082
V202	1227	Aug 1941	Nov 1946	Surplus. N68361
V203	1228	Aug 1941	Nov 1946	Surplus. N1340V
V204	1229	Aug 1941		
V205	1253	Feb 1942	Aug 1946	
V206	1254	Feb 1942		
V207	1255	Feb 1942		
V208	1256	Mar 1942		
V209	1257	Mar 1942		
V210	1258	Mar 1942	Nov 1946	
V211	1259			
V212	1260	Apr 1942	Apr 1948	Surplus. N743 N2770A N324BC
V213	1261	May 1942	Nov 1946	Surplus. N5508N
V214	1262			Surplus. N700A
V215	1263			
V216	1264	May 1942		
V217	1265	Jun 1942	Mar 1947	Surplus. N1284N CF-IIQ
V218	1266	Jun 1942	Apr 1948	Surplus. N744 N151M
V219	1267	Jun 1942	Dec 1946	Surplus. N90140
V220	1268	June 1942	Dec 1946	Surplus. N91039
V221	1269	Oct 1942		

Notes: V212 and V218 initially went to the US Department of the Interior.

GRUMMAN UF-1G/2G HU-16E ALBATROSS

Defined in Webster's dictionary as the largest of the web-footed sea birds, is the Albatross. Grumman developed a successful amphibian after World War II and named it Albatross. Analogous to the living bird, Grumman's aluminium creation is the largest of their amphibious aircraft family. An unusual military aircraft, the Albatross was developed by Grumman as a large amphibian replacement for their Goose, Widgeon and Mallard.

Grumman believed in the practicability of amphibians over a longer period of time than any other aircraft manufacturer, and the Coast Guard were the first to recognise the Albatross as a replacement for their Goose and Widgeon fleet. Initially the Coast Guard and the US Navy lacked appropriation of funds to purchase in any large quantity. A prototype was in fact ordered by the US Navy. It was left to the US Air Force who were looking for a versatile search and rescue (SAR) aircraft to replace aging make-shift converted Boeing B-17 and B-29 bombers as search aircraft. Admittedly, they both carried a lifeboat attached to the lower fuselage. This could be dropped to survivors, but this method of rescue was not the immediate recovery of the personnel. The US Air Force contracted to purchase 290 Grumman Albatross amphibians designated SA-16A.

In the early 1950s the Coast Guard aircraft inventory contained a variety of vintage and veteran World War II types which it was keen to reduce and consolidate on a few types especially in the search and rescue (SAR) role. The miscellaneous air fleet consisted of Boeing PB-1Gs, Douglas R4D-5s, Consolidated PBY-5/6G Catalinas, and Martin P5M Mariners.

When the US Navy selected the Grumman G-64 designation XJRF-1 was assigned to the prototype. The 'JR' indicated a utility transport, and JR2F-1 was the first of no less than five successive designations applied to this US Navy amphibian over the years. The prototype flew on 24 October 1947 and evaluation flying being successful the US Navy confirmed its proposed order, and prompted the US Air Force interest in the type who ordered it in quantity. The US Navy abandoned the 'JR' designation in favour of 'PF-1' in the patrol category, and it then became 'UF' in the US Navy's new utility series. The first Grumman UF-1s were similar in most respects to the prototype.

The US Coast Guard received more than eighty-eight Albatross amphibians over the years. The thirty-one Grumman UF-1Gs ordered closely matched the specification of the US Navy UF-1 amphibians. Fifteen ordered by the US Air Force were transferred to the US Coast Guard, being renumbered before completion. These were followed by thirty-seven SA-16A aircraft which were transferred after first seeing service with the US Air Force. In Coast Guard service, these latter aircraft used serial numbers corresponding to their US Air Force serial numbers without the first digit and the hyphen. The UF-1G was ordered during April 1950, and was first flown in May 1951. All but five of the USCG UF-1Gs were later returned to Grumman at Bethpage and brought up to UF-2G standard. Three more Albatross amphibians were transferred from the US Air Force in December 1953, five more in January 1954, and seven the following month in February. The Albatross 2121 ex-52-121 crashed in December 1954.

HISTORY

During 1944 design work commenced on a general-purpose amphibian for the US Navy, as a successor to the JRF Goose which served well throughout World War II. The Grumman company had the benefit of more than ten years' experience in the design and production of amphibians for both the US Coast Guard and the US Navy. The Model G-64, as the new type was designated, was a worthy continuation of the JRF Goose design philosophy. If featured a convential two-step hull into the sides

Over eighty Grumman Albatross amphibians served faithfully with the USCG over the years, and it was a true workhorse with the service. Depicted is an early UF-1G '1276' delivered in November 1953. Later models were the improved UF-2G, some UF-1Gs being returned to Grumman for modification. *(USCG)*

of which the main wheels retracted, had a high wing with fixed stabilising floats and a single tail unit.

When the US Air Force was established on 18 September 1947 it was soon given responsibility for world-wide air rescue operations, a role it shared with the US Coast Guard. For Grumman it was the gain of a new primary customer, the US Air Force, for the Design 64, initiated during April 1944 for the US Navy. Evolved as a replacement for the JRF Goose utility amphibian, the new design was of similar configuration, being twin-engined, but larger and more powerful than its predecessor, and intended for all-weather operations. It proved to have better water and land-handling charcteristics, the former being improved as the result of work undertaken by Ralston Stalb, a noted hydrodynamicist, and the latter coming about through the adoption of a tricycle undercarriage with a wider track main gears attached beneath the engine nacelles and retracting into the fuselage sides.

Ordered by the US Navy in November 1944, it assigned the designation XJR2F-1 to the two prototypes and named it Pelican. However, production versions of the G-64 amphibian were named Albatross and its production history is the most complex of any of the Grumman types. In numerous instances, aircraft which had been ordered by one branch of the US armed services, and allocated appropriate designation and serials, were delivered to another branch and assigned new designation and serials as required. In other cases, aircraft were given new designations and serials upon being transferred from one branch of the armed services to another. A good example is c/n G-88 delivered to the US Air Force as an SA-16A 51-015, later transferred to the US Coast Guard to be designated UF-1G and numbered 1015.

Following US Air Force practice with its SA-16A amphibians, the US Navy and US Coast Guard initiated conversion of its early UF-1s to UF-2 standard in 1957. Among the several modifications were an increase of 16½ft in the wing span, use of fixed leading-edge camber in place of slots, enlarged ailerons and an extended taller fin and rudder. The Coast Guard had a total of thirty-four UF-1G aircraft converted to UF-2G standard and, as mentioned earlier, received a further thirty-seven from the US Air Force, these eventually being designated HU-16E.

The US Coast Guard received fifteen Albatross amphibians ordered by the US Air Force before completion at Bethpage, and already assigned USAF serials 52-121 to 52-135 and delivered to the USCG as UF-1Gs with USCG serials 2121 to 2135 adapted from the original USAF serial. Power was from two Wright 1425hp R-1820-76A engines. For short take-off in open sea conditions, or from short air strips, it could be fitted with JATO bottles strapped on each side of the aft fuselage. Normal crew consisted of pilot, co-pilot, navigator and radio operator, with two flight observers being added during search and rescue

(SAR) sorties. The main cabin could be configured to carry either ten passengers or twelve stretcher patients plus one attendant, or passengers and litter patients. External store racks were fitted to each wing, available for either 295 US gallon drop tanks, or rescue kits and supplies up to 2000lb. Some Albatross aircraft carried the AN/APS-31A search radar, initially fitted in a nacelle beneath the port wing, later in the nose. The Grumman design numbers G-211, G-234, G-270 and G-288 used were dependant on equipment changes and systems improvements.

The Achilles' heel of the early Grumman UF-1G Albatross, which led to a number of accidents, was traced to the wing design.

The original production aircraft had one extremely undesirable characteristic. Single-engine operation consisted of a controlled descent — and some were more controlled than others. The Grumman engineers got to the bottom of the problem by extending the wing from eighty feet to ninety-six feet eight inches. This modification resulted in a vast improvement in

This Grumman UF-2G is in fact an ex USAF SA-16B Albatross 51-7240 transferred with over fifty others to the USCG. This aircraft '7240' is seen in 1958 low flying over rough seas from its base at Argentia, Newfoundland. During 1959 it assisted in the 1959 International Ice Patrol in the Atlantic. *(USCG)*

performance, since single-engine flight at altitude was made possible. Maximum gross weight was increased by 5000lb; cruise airspeed was increased by more than twenty-five knots with no increase in fuel consumption; stall speed was lowered from seventy to sixty-four knots, and the maximum range was extended some 500 miles. The Coast Guard converted seventy-eight of their early models to the new stretched-wing version designated UF-2G.

During May 1951 the first USCG UF-1G 1240 was delivered from the Grumman factory; the last Grumman HU-16E 7250 ex-US Air Force SA-16A 51-7250 was retired at Cape Cod air station, Massachusetts during March 1983. Under a US Department of Defense Directive dated 6 July 1962 all designations allocated to US Navy and US Coast Guard aircraft were standardised with those of the many US Air Force variants. In the intervening years, the Coast Guard Albatross flew rescue, fishery patrol and pollution surveillance missions from air stations located in Alaska — Annette and Kodiak; California — San Diego, San Francisco and Sacramento; Florida — Miami and St Petersburg; Hawaii — Barbers Point; Massachusetts — Cape Cod; Michigan — Traverse City; Mississippi — Biloxi; Oregon — Salem; Texas — Corpus Christi; Puerto Rico — San Juan; Washington — Port Angles, and the Philippines from Sangley Point.

In addition to their regular duties these amphibians were often flown as utility transports for the benefit of various US government agencies and VIPs. One Coast Guard HU-16E '7246' was modified in 1975 to test the new Airborne Oil Spill Surveillance (AOSS) system fitted with a sideways-looking airborne radar (SLAR) housed in a fairing at the rear of the starboard hull side and with other pollution detection system equipment mounted in the bow, in a nacelle attached beneath the cockpit on the port side, and in a modified external tank fitted beneath the starboard wing. The two outer wing floats were removed and the aircraft went to Edwards Air Force Base test centre in California to determine the effects of the SLAR antenna and the other extensive modifications on the aircraft performance. For these extensive flight tests the complete rear of the HU-16E fuselage and tail unit were tufted.

Initially the test aircraft was operated from the USCG air station at Miami, Florida, later moved across the United States to San Francisco air station where its capability of detecting pollutents could be more effectively used. The Albatross carried the system manufacturer's motif and the name *SeaVeyor*.

THE GOAT

For many years, the unofficial nickname used by the Coast Guard for the Albatross has been the 'Goat'. No one is quite sure where the name originated, but one story relates that when the Coast Guard first acquired the Lockheed C-130 Hercules, the

Seen on 15 January 1969 is HU-16E Albatross '7242' ex USAF 51-7242 in the latest USCG livery in use today with its undercarriage about to retract into the fuselage housing. The last operational sortie of the *Goat* was made on 10 March 1983 from Cape Cod Air Station, Massachusetts. *(USCG)*

new C-130 pilots did not want to be associated with a 'smelly old goat'. Another yarn tells how Federal Aviation Administration tower operator at Washington National Airport thought that a HU-16E parked in the grass behind the Coast Guard hangar looked like an 'old goat' grazing in the pasture.

Other Coast Guard pilots developed another tongue-in-cheek name for the rather noisy Grumman amphibian. Because of the deafening roar at take-off and loud drone at cruise, the name 'Whispering Goat' was quoted. Due to the relatively short propellers required for operation from water, an extremely loud, high-pitched roar occurs at maximum power. The engines were Wright Cyclone R-1820-76 which developed 1425bhp at take-off. The sound was sufficiently unique that it was possible to differentiate a HU-16E taking-off from that of other Wright 1820-powered aircraft such as the Douglas DC-3, Lockheed Lodestar, North American T-28, or Grumman S-2 Tracker. With JATO bottles fitted for take-off even more noise was generated by the Albatross, and it was sometimes remarked that a JATO departure by a 'Goat' was noiser than that of any other reciprocal or jet aircraft then in existance.

Sheltered water take-offs and landings at weights up to 32,000lb were possible without the use of JATO bottles. Open sea operations were possible under favourable conditions with JATO bottles and there has been at least one recorded take-off of an Albatross without JATO from seas as high as seven feet. The tricycle landing gear permitted normal land operation. It was one of the largest aircraft not to employ a steerable nose wheel. The nose wheel castors freely with differential power, rudder, brakes, and reverse thrust used for directional control. With the addition of four 1000lb thrust JATO bottles, both land and water take-off distances were reduced by forty per cent. The Hamilton Standard reversible-pitch propellers enabled the amphibian to be stopped in less than 900 feet. This feature was particularly advantageous on rough water to reduce a pounding roll-out. The all-weather features of the Grumman enabled the Albatross to fly under stormy winter conditions. Anti-ice and de-ice systems kept the airframe relatively free of ice accumulations.

A series of cargo tie-down points enabled the HU-16E to be used for carrying cargo. This was normally loaded through the aft door hatch, but a large overhead hatch permitted loading of packages up to five feet across. This allowed a replacement engine to be airlifted to a grounded aircraft, and with little attention to fuel loading, up to 5000lb of cargo could be carried. Avionic packages were available for search and rescue (SAR) missions. Radar, HF SSB transceivers, interrogrators, and MF/VHF/UHF direction-finding equipment was standard on all Coast Guard aircraft for the aid in the location of disabled vessels, ditched aircraft, and isolated liferafts with survivors. Fuel was located in three tanks — internal wing, droppable bomb-type tanks, and wing floats. The capacity of 10,086lb of fuel permitted an endurance of almost twenty-four hours for long range deployment if required.

OPERATIONS

A total of 466 Albatross amphibians were built by Grumman between September 1947 and May 1961. Grumman operated an overhaul and modification facility for the Albatross at Stuart, Florida.

There are many examples of how the Coast Guard Albatross was adapted under most trying circumstances. South-eastern Alaska is extremely mountainous, and the shoreline is composed of treacherous fjords, not unlike Scandinavia. Both winter and spring weather in that part of Alaska is very severe. Very low ceilings and visibility coupled with frequent rain and ice storms, are all characteristics of that area.

During late winter 1964 the Gran Doc Mine thirty-five miles east of Ketchikan, Alaska, was enveloped by an avalanche which buried all sixty miners beneath tons of snow. Four days after the disaster a Coast Guard HH-52A helicopter accompanied by a civilian helicopter, commenced rescue operations, with a cloud base of less than 200 feet. A Grumman HU-16E from Coast Guard air station located at Annette, piloted by Lieutenant Ed Murnane, flew by radar through the narrow winding fjord to evacuate the survivors the two helicopters had removed from the mountain side. No suitable temporary runway was available near the mine, so the Albatross was forced to land in the bay, accompanied by a gusting 40-knot wind and water temperature of minus 35 degrees. Three round trips through the needle-like passageway flying at fifty feet were necessary to remove all the miners. It was a miracle that all survivors reached safety, thanks to the superhuman effort of the Coast Guard crew and the 'Goat'.

'Now put the ready HU-16 on the line — fishing boat sinking 150 miles south-east of Miami.' Such terse loudspeaker announcements accompanied by the 'beep, beep, beep' of the scramble alarm klaxon has been heard hundreds, perhaps thousands of times over the public address system, not only at Miami but at any Coast Guard air station or air facility during the long lifespan of the Grumman HU-16E.

Even after restrictions were placed on the 'Goat' which forbade water landings, the type was often called upon to act as 'On Scene Commander', controlling at times the activities of as many as ten aircraft and ships provided by all of the military services, the Civil Air Patrol and the volunteer Bahamian Air Sea Rescue Service — BASRA.

The crash of the Eastern Air Lines Lockheed L-1011 in the Florida Everglades in December 1972 is a foremost example of the communications versatility of the Albatross. The aircraft directed the rescue efforts of four Coast Guard and two US Air Force helicopters, arranged coordination with Miami Approach Control, vectored the rescue and news media helicopters around each other at the crash site, and gave updates and progress reports on the rescue to the Miami Coast Guard Rescue Coordination Centre.

Frequently, small boats would break down in the Gulf Stream resulting in large scale searches. On one occasion, five HU-16s, five HU-52s and HH-3F helicopters, plus three Coast Guard cutters, were assigned to search for a sixteen-foot outboard motorboat. Many such searches lasted four or even five days, and on occasion up to eight days if there was the slightest hope of finding survivor(s).

During 1972, the Grumman amphibian participated in searches for approximately 400 boats, liferafts, and aircraft in the Caribbean area. While normal searches were conducted 300 miles from Miami, some occurred as far away as Antigua and Honduras, about 1000 miles from the home station located at Opa Locka aiport. Medical evacuations, or medevacs, were an important lifesaving mission which involved the 'Goat'. Warm, clear water and coral reefs today attract many divers to the sub-tropical latitudes. Several times each year, a HU-16 would be dispatched to bring home a diver, suffering from the bends, to a US Navy decompression chamber in either Fort Lauderdale or Key West. Once picked up the victim had to be flown at a fifty foot altitude at maximum speed to prevent further nitrogen damage. There are many isolated islands with insignificant or no air service. It was the Coast Guard Albatross that was often called upon to transport accident victims, appendicitis patients and heart attack cases to mainland hospitals. Airevacs operated by the service have saved or aided many over the years.

Smuggling, which used to be thought of as rum running, is now primarily 'grass running' or drug smuggling. Before retirement the HU-16s located many illicit operations and prevented many tons of illegal drugs from being sold on the US market. The Albatross conducted patrols to survey foreign fishing vessels operating in US territorial waters, and even kept an eye on American ships fishing near foreign countries. The ecology movement in the United States has made everyone more aware of conservation and pollution control. Several modified Grumman HU-16s were used by the Coast Guard to carry infra-red and ultraviolet equipment to detect and photograph oil slicks in harbours and shipping lanes. Much of the US coastline is patrolled several times a week to discourage ships, including tankers, from pumping bilges and dumping other pollutants into the water.

With the advent of the Coast Guard helicopter rescue activities by the 'Goat' diminished, but far out at sea the HU-16 was still required on task to drop food, water, radios and rafts to survivors. Likewise, leaking boats could be supplied with dewatering pumps by the ubiquitous Albatross in order to contain flooding. In their time they have flown many miles out into the ocean to escort aircraft in trouble to a safe landing. In 1973, a Piper Apache ran out of fuel east of Jacksonville, Florida, and it was a Miami-based Coast Guard HU-16E which provided ditching instructions.

All the early UF-1G and SA-16A models that were converted to UF-2G and SA-16B standard had a series of holes drilled in the wing spar as part of the wing lengthening programme. A test at the US Navy research centre at Patuxent River, Maryland, revealed that catastrophic failure along the holes might result at 19,000 hours flight time. Accordingly, a flight restriction of 11,000 hours was placed on all Albatross airframes. However, twenty-one of the type newly constructed in 1959 as later models did not have the holes drilled, so the definite service life was never established.

Most US Air Force, US Navy and US Coast Guard aircraft are stored at the huge storage depot or 'boneyard' located at Davis-Monthan AFB in Arizona. Many of the 'Goats' in the Coast Guard inventory made their last ferry flight to the Arizona boneyard, although a number have been preserved in museums and elsewhere. Water landings for the HU-16E were terminated in 1972, but she still continued in service.

During 1966 the Coast Guard elected to participate jointly with the US Navy, US Air Force and the Royal Canadian Air Force in a full-scale wing fatigue test of the Grumman HU-16E — in fact it was the US Coast Guard who provided the airframe to the US Naval Air Development Center located at Warminster,

Pensylvannia, for the testing. The object of the test was to determine whether major wing repair or replacement of the wing should be required. Testing commenced on 24 January 1968 and, after a myriad of systems and funding problems, terminated on 31 October 1968 with the failure of the starboard wing. Subsequent engineering calculations resulted in the establishment of a wing service life of 11,000 flight hours for the USCG HU-16E Albatross fleet. When the wing fatigue testing was initiated, the HU-16E fleet numbered seventy-six aircraft, down three from the 1963 total of seventy-nine.

RECORD BREAKERS

A glance at the HU-16E does not inspire thoughts of speed and power; however, the Albatross is the holder of nine world class amphibian records, certified by the Federation Aeronautique International (FAI), the world's governing body for aviation records. The records were established as a result of a tri-service venture by the Coast Guard, US Navy and US Air Force and are still valid. Moreover, unless interest in amphibian aircraft development is renewed these records may stand forever. Records established by Grumman HU-16 Albatross 7255 are as follows:

1, 2 Speed over a 1000-km closed course with a 1000kg and 2000kg load. Established by Commander Wallace C Dahlgreen and Commander William G Fenlon on 13 August 1962, at a speed of 201.5 knots.

3 Speed over a 5000-km closed course with a 1000kg load. Established on 15 and 16 September 1962 by two US Navy pilots flying the Coast Guard aircraft at a speed of 131.5 knots.

4, 5 Altitude with a 1000kg load and altitude with a 2000kg load. Established on 12 September 1962 by two US Navy pilots flying the Coast Guard aircraft at altitudes of 29,475ft and 27,405ft respectively.

6 Distance — non-stop. Established on 24 October 1962 on a flight from USCG air station Kodiak, Alaska, to the US Naval Air Station at Pensacola, Florida, a distance of 3104 nautical miles, by Commander William G Fenlon, Commander Wallace G Dahlgreen, Lieutenant W Senn and Chief W Taggart, USCG.

When the aircraft carrier USS *Intrepid* commenced her second career as a historical exhibit in New York, Coast Guard Aviation jumped at the opportunity to place a Grumman HU-16E Albatross aboard. The problem was that the amphibian 7216 had been sitting outside for ten years and needed major rework, inside and out. Starting in April 1983 the Coast Guard Auxiliary from the 3rd Northern District got down to the job and after more than a year of hard work, the amphibian was completed and hoisted aboard. The restoration project was headed by

Lieutenant Ed Ward of the Coast Guard Auxiliary who holds both the FAA Airline Transport and the Airframe and Powerplant Mechanics ratings. Volunteers came from the USCG air station located at Brooklyn, New York.

The last operational Grumman HU-16E Albatross 7250 made its final flight on 10 March 1983 at the USCG air station Cape Cod, Massachusetts. The crew consisted of Commander Eric J Stout, pilot, Lieutenant David E Elliott, co-pilot, ADCM John E Bloom, CWO Dean A Long and CWO Stephen P Marvin, crewmen. Commander Stout had logged over 2800 hours in the HU-16E. At the farewell ceremony Rear Admiral Louis L Zumstein, Commander First Coast Guard District, said, 'I've flown forty-six HU-16s in my career and they have brought me home when the odds were stacked against me several times'. For more than 500,000 flight hours, it performed a variety of Coast Guard missions. The Albatross was used extensively during the Cuban boat exodus in the 1960s and again in the 1980s. Thousands of Cuban refugees owe their lives to the men who flew the 'Goats', patrolling the waters between Cuba and Florida. Today 7250 is on display at USCG air station Cape Cod, a monument not only to all those who flew the Albatross, but to US Coast Guard Aviation.

TECHNICAL DATA

Manufacturer:	Grumman	Prop diameter:	11 ft
Other designations:	SA-16A	Designation:	UF-1G
Contract No.:	10988 51-118 51-656	Type:	General purpose amphibian
Span:	80 ft	Unit cost:	$321,000
Height:	24 ft 5 in	Length:	60 ft 7 in
Fuel:	600 galls 676 with drop tanks	Wing area:	833 sq ft
Empty weight:	19,820 lb	Oil:	58 galls
Crew:	2	Gross weight:	33,000 lb
Top speed:	240 mph at 7600 ft	Passengers:	6
Cruise:	150 mph at 1500 ft	Stall speed:	88 mph
Sea-level climb:	1090 ft per min	Range:	2660 miles at 1800 ft
Engine:	Wright Cyclone R-1820-76A	Service ceiling:	31,300 ft
Take-off power:	2 × 1425 hp	Three-blade prop blades	Hydromatic 600 1A-7

Miscellaneous information:
Contracts: 53-394.

TECHNICAL DATA

Manufacturer:	Grumman	Prop diameter:	11 ft
Other designations:	SA-16B	Designation:	UF-2G, HU-16E
Span:	96 ft 8 in	Type:	General purpose amphibian
Height:	25 ft 10 in		
Empty weight:	22,883 lb	Length:	62 ft 10 in
Crew:	4	Wing area:	1035 sq ft
Top speed:	236 mph	Gross weight:	30,353 lb
Cruise:	171 mph	Passengers:	6, 10 stretchers
Sea-level climb:	1170 ft per min	Range:	1715 miles
Engine:	Wright Cyclone R-1820-76A/B	Service ceiling:	23,500 ft
		Prop blades	Hydromatic 600 1A-7
Take-off power:	2 × 1425 hp		

SERIAL INFORMATION

USCG No	Contract	Commissioned	Decommissioned
1240	NOa(s) 10988	May 1951	
1241		Jun 1951	
1242		Aug 1951	
1243		Sept 1951	
1259	NOa(s) 51-118	Mar 1952	
1260		Apr 1952	
1261		May 1952	
1262		Mar 1952	
1263		May 1952	
1264		Jun 1952	
1265		Jun 1952	
1266		Aug 1952	
1267		Aug 1952	
1271	NOa(s) 51-656	Mar 1953	
1272		Apr 1953	
1273		May 1953	
1274		Jun 1953	
1275		Oct 1953	
1276		Nov 1953	
1277		Nov 1953	
1278		Dec 1953	
1279		Dec 1953	
1280	NOa(s) 53-394	Jan 1954	
1288		Apr 1954	
1289		Apr 1954	
1290		Apr 1954	
1291		Apr 1954	
1292		May 1954	
1293		May 1954	
1294		May 1954	
1311			
1313			
1314			
1315			
1316			
1317			

SERIAL INFORMATION

USCG No	Commissioned	Decommissioned
	Grumman HU-15E Albatross ex-US Air Force	
51-015	51-7243	
51-016	51-7245	
51-023	51-7246	
51-026	51-7247	
51-030	51-7248	
51-7188	51-7250	
51-7209	51-7251	
51-7213	51-7254	
51-7214	51-7255	
51-7215	52-121	
51-7216	52-123	
51-7218	52-124	
51-7223	52-125	
51-7226	52-126	
51-7227	52-127	
51-7228	52-128	
51-7229	52-129	
51-7230	52-130	
51-7232	52-131	
51-7233	52-132	
51-7234	52-133	
51-7236	52-134	
51-7237	52-135	Total number of Albatross
51-7238		aircraft operated by the US
51-7239		Coast Guard is quoted as 88. the
51-7240		total produced here is 87.
51-7241		
51-7242		

GRUMMAN VC-4A GULFSTREAM I

Grumman, in search for a successor to its long line of twin-engined transport amphibians, investigated a broad range of possibilities prior to undertaking the development of their Design 159, an all-new propeller turbine powered aircraft. On 19 March 1963, the US Coast Guard accepted a single Grumman G-159 Gulfstream I which was designated by the Department of Defense as VC-4A.

Accepted by the USCG on 19 March 1963, this Grumman VC-4A Gulfstream I executive aircraft is seen on 28 March 1973, from its base at Washington National Airport. Initially it was '1380' in the USCG inventory, now '02' as part of the VIP fleet used by the USCG Commandant and others. *(USCG)*

The new aircraft soon proved to be ideally suited for the air transportation mission and gained immediate acceptance by crew and passengers alike. It was based at the US Coast Guard Arlington air station located at Washington National Airport. During February 1974 the unit moved to share a facility with the Federal Aviation Administration in Hangar 6 and the air station became 'Washington' by name. The Gulfstream I was utilised by both the Secretary of Transportation and the US Coast Guard Commandant.

The VC-4A was the ninety-first Gulfstream I and was initially assigned the USCG serial 1380 later becoming 02, Coast Guard Zero Two. A second VC-4A was ordered and allocated the USCG serial 1381 but was later cancelled. During September 1983 the Gulfstream I 02 was re-assigned to the USCG air station located at Elizabeth City and it is understood it has corrosion problems. It has the Grumman c/n 91.

Between August 1958 and May 1969 Grumman built no less than 200 Gulfstream I aircraft. In the United States the type has been used by five government agencies and branches of the US armed forces. Power is provided by two Rolls-Royce Dart 529 turbines driving four-blade propellers and with a crew of two the Gulfstream I can carry between ten and fourteen passengers. No major modifications were necessary during the course of production.

TECHNICAL DATA

Manufacturer:	Grumman	Designation:	VC-4A Gulfstream I
Other designations:	G-159	Type:	Executive transport
Span:	78 ft 4 in	Length:	64 ft 6 in
Height:	22 ft 9 in	Wing area:	615 sq ft
Fuel:	1550 US galls	Oil:	7 galls
Empty weight:	21,900 lb	Gross weight:	35,100 lb
Crew:	2	Passengers:	10/14
Top speed:	348 mph at 25,000 ft	Stall speed:	128 mph
Cruise:	288 mph at 25,000 ft	Range:	2540 miles
Sea-level climb:	1900 ft per min	Service ceiling:	26,000 ft
Engine:	2 × Rolls-Royce Dart 520 turboprop	Four blade prop blades	Rotol constant speed

GRUMMAN VC-IIA GULFSTREAM II

During February 1969, a single Grumman Gulfstream II executive transport was delivered to the US Coast Guard. Increasing worldwide travel requirements by both the Secretary of Transportation and the US Coast Guard Commandant provided ample justification for the acquisition of a new high-speed executive jet transport. Given the designation VC-11A and the USCG serial 01 it entered service as 'Queen of the Fleet'. It is normally based at the USCG air station located at Washington National Airport, District of Columbia, previously known as Arlington air station.

Purchased off-the-shelf it is powered by two Rolls-Royce Spey Mk 511-8 (RB 163-25) turbofans, giving a maximum take-off power of 11,400lb thrust and maximum continuous power delivering 10,940lb of thrust. The VC-11A carries two wing tanks with a cross-flow line which allows the fuel to flow from tank to the opposite engine or both engines. Each tank contains a reservoir of 190 gallons in separate compartments at the wing centre. Usable fuel is 3450 gallons or 23,300lb. The Gulfstream II has a maximum cruising speed of 511 knots and a maximum range of 2930 nautical miles. Maximum take-off weight is 59,500lb. It can operate up to altitudes of 43,000ft.

Precision navigation to any point in the world without outside input is possible through use of the aircraft's Inertial Navigation System (INS). The aircraft is manned by a crew of four and can carry a maximum of twelve passengers. At a ceremony held at the Washington air station early in 1988 the 10,000th flight hour was celebrated. Until the introduction of the Falcon Jet HU-25A in the early 1980s 'Coast Guard Zero One' was the only jet fixed-wing aircraft in US Coast Guard service.

Following the debut of business jets such as the Lockheed Jetstar and North American Sabreliner, which respectively obtained their FAA Type Approvals in August 1961 and March 1962, Grumman decided that if it was to stay in the executive aircraft market it would have to find a successor to the successful Gulfstream I which was powered by propeller turbines. Existing and prospective customers were contacted, with the result that there was a need for a jet-powered successor to the Gulfstream I.

The resulting Design 1159 retained much of the fuselage of the earlier Design 159 and was also fitted with dual nose and main-wheels. New wings, with 25 degrees of sweep at quarter-chord, and swept T-tail surfaces were designed. The specification and performance figures were circulated to prospective customers in late 1964 and early 1965. The result was sufficient orders to announce a firm commitment for the Gulfstream II programme on 5 May 1965.

First seven Gulfstream IIs were built in Bethpage, c/n 008 becoming the first to be completed in Georgia in the new

Now the flagship with the USCG is '01', a Grumman VC-11A Gulfstream II delivered during February 1969, and based at Washington National Airport. It is powered by Rolls-Royce Spey engines, and this high speed executive jet transport travels worldwide. Photo was taken on 23 March 1973. *(USCG)*

purpose-built plant leased from the City of Savannah and the Savannah Airport Commission. Very few modifications were introduced during production, those of any significance being the installation of hush kits which enabled the aircraft to be certificated under new FAR Part 28 and Part 36 regulations, the fitting of tip tanks increasing fuel capacity from 3585 to 4123 US gallons and extending the range by 400 nautical miles. Lastly, the factory updated the avionics.

All but one of the Gulfstream IIs were built for private customers or for civil government agencies in the United States and abroad. The exception was c/n 023 which was completed as a VIP staff transport for the US Coast Guard. According to the Grumman history 'Coast Guard Zero One' was delivered during July 1968.

TECHNICAL DATA

Manufacturer:	Grumman	Designation:	VC-11A Gulfstream II
Span:	68 ft 10 in	Type:	Executive transport
Height:	71 ft 4 in	Length:	79 ft 11 in
Fuel:	3450 US galls	Wing area:	793.5 sq ft
Crew:	4	Gross weight:	59,500 lb
Top speed:	585 mph	Passengers:	19
Cruise:	565 mph	Stall speed:	154 mph
Sea-level climb:	5050 ft per min	Range:	2930 nm
Engine:	2 × Rolls-Royce Spey Mk 511.8 turbofan	Service ceiling:	43,000 ft

GRUMMAN E-2C HAWKEYE

In an effort to stop the flow of illegal drugs into the United States, the Coast Guard is using the Grumman E-2C Hawkeye surveillance aircraft. These airborne early-warning pickets have a key role to play in the new mission of the Coast Guard of air interdiction. The E-2C allows the USCG to survey nearly three million cubic miles of surrounding air space at sea, while monitoring 150,000 square miles of ocean surface. As an example, an E-2C Hawkeye, flying over New York City can monitor all the air traffic in the heavily congested Boston to Washington, DC, corridor while maintaining a watch over surface vessels in the same Atlantic ocean area.

Airborne early warning (AEW) radar systems to detect targets beyond the line of sight of surface ships, after continuous development by the US Navy, led in 1956 to a new concept of US Naval Tactical Data System. It was designed to provide a task force commander with all the information on the disposition of both ships and aircraft — friend and foe — needed to control his forces. Playing a key part in this concept was the airborne early-warning picket aircraft, carrying in addition to long-range search radar, digital computers which would automatically detect targets and select the best available interception available to the despatched to meet each target.

The US Navy requirement was made the subject of a US industry-wide design competition which was won by the Grumman Aerospace Corporation of Bathpage, Long Island, New York, on 5 March 1957, with the G-123 design, later named the Hawkeye and designated initially W2F-1. It was a high-wing monoplane with the stacked antennae elements located in the twenty-four foot diameter rotating dome above the fuselage. Power was provided by two Allison T56-A-8 turboprops. Due to the peculiar airflow over and around the enormous radome Grumman had to use a multiple-surface tail unit.

Making its first flight on 21 October 1960 the W2F-1 designation was subsequently changed to E-2. The prototype did not have the full Hawkeye electronic systems, these becoming airborne in a fully-equipped E-2 on 29 April 1961. Deliveries commenced to US Navy units on 19 January 1964. The E-2B was an improved version of the E-2A, the principal difference being a Litton Industries L304 micro-electric general-purpose computer.

The Hawkeye's operational capability was improved greatly with the introduction of the E-2C, the first prototype flying on 20 January 1971, followed by a second a year later. Both were conversions of the earlier E-2A. The avionics system of the E-2C was completely revised, with an AN/APA-171 antenna system in the rotating radome and an AN/APS-120 search radar fitted in the nose. In addition to the Litton Industries L-304 computer as used in the E-2B, the E-2C had an air data computer, a carrier aircraft inertial navigation system (CAINS) and uprated engines.

The first production E-2C flew on 23 September 1972 and the type entered service with the US Navy in November 1973.

Production of the E-2C by Grumman continued at a low annual rate to build up the US Navy's AEW force to no less than fourteen squadrons by the late 1980s. Orders totalled 144 by 1988, plus export contracts. Four E-2Cs were transferred by the US Navy for use by the US Coast Guard and the US Customs Service for anti-drug running surveillance. The Hawkeye can use short runways, has a low search speed, and long extended mission time. For better storage capability, the wings of the E-2C Hawkeye fold upwards.

CGAW-1's MISSION

The Anti-Drug Abuse Act of 1986 established a US Coast Guard role in the interdiction of air smuggling. This legislation increased the resources available to the Coast Guard to help stop illegal substances destined for the United States via the air routes over the maritime region. The legislation included authorisation for the US Navy to initially loan two Grumman E-2C Hawkeye aircraft for use in combating drugs brought into the US by aircraft.

Still in US Navy livery but with USCG serial '3502' on the nose is Grumman E-2C Hawkeye BuNo 159497 from CGAW-1 the Coast Guard Airborne Warning Squadron. Operated successfully by the US Navy, the E-2C is involved in the effort to stop the flow of illegal drugs into the United States. *(Grumman)*

The USCG air facility located at Norfolk, Virginia, was formed with two E-2C Hawkeye aircraft on 22 January 1987, initially operating out of the US Navy CCAEWW-12 headquarters building — Commander, Carrier Airborne Early Warning Wing 12 — and as the unit was home-based at a US Naval Air Station it was necessary to attach a designation which fitted the US Navy operation procedures. That designation was CGAW-1 — Coast Guard Airborne Warning Squadron One. The first Commanding Officer was Commander Norman V Scurria Jr.

Carrying the Coast Guard serials 3501 and 3502 the first E-2C aircraft are interesting in that 3501 was the 45th production aircraft with the US Navy BuNo 160698, while 3502 is the 23rd production E-2C having the US Navy BuNo 159497. The aircraft deploy as required to the Caribbean, Gulf of Mexico and off the Atlantic ocean coastline to participate in detecting suspect aircraft operated by the drug barons.

The E-2C Hawkeye is one of the most capable searchers used by the US Coast Guard. Fitted with the AN/APS-125 radar, they fly nearly 1000 hours a year each. Two that were loaned by the US Navy to the US Customs Service were returned to the US Navy in the summer of 1989 and then added to the strength of CGAW-1. In July 1989 a new E-2C base for the Coast Guard aircraft was opened at the Grumman facility located at St Augustine, Florida. It proved beneficial in more ways than one, and aircraft can easily be deployed to such Caribbean vantage points in the anti-drug war as Key West, Florida.

The introduction of the Grumman E-2C Hawkeye aircraft, and the future introduction of a Coast Guard Lockheed HC-130H equipped with a Hawkeye-type radar on top of the fuselage for early warning, has produced a new aviation task and a new insignia, these being the Coast Guard Flight Officer (CGFO) holding rank as commissioned officers and in the new unit operating the radar systems. Their insignia is identical to the US Navy Flight Officer's wings, and CGAW-1 was the first Coast Guard unit to have flight officers. In addition to the CGFOs, the unit has also enlisted radio operators called Flight Technicians (FLTTECHS). These are Avionics Technicians (AT) who have gone through specialised training and are qualified to operate the radar equipment in the E-2C. They wear the aircrew insignia which all of the Coast Guard's enlisted aircrew are entitled to. Flight crews for the Grumman E-2C are normally made up of two pilots and three radar operators. There is at least one CGFO on board who is in charge of the radar operations.

The radar barrier imposed by the patrolling E-2C Hawkeye aircraft is bolstered by US Coast Guard Aerostat units. Five are ship-based plus one at High Rock, Bahamas, call-sign 'Cariball One' while 'Cariball Two' is located on the abandoned Georgetown airport at Great Exuma, Bahamas. 'Cariball three' has been proposed, if funds are appropriated, and will operate from Great Inagua, in the southern Bahamas.

TECHNICAL DATA

Manufacturer:	Grumman	Take-off power:	2 × 4910 shp
Other designations:	G-123	Designation:	E-2C
Span:	80 ft 7 in	Type:	Airborne early warning picket
Height:	18 ft 4 in	Length:	57 ft 8 in
Fuel:	12,400 lb	Wing area:	700 sq ft
Empty weight:	38,867 lb	Gross weight:	52,730 lb
Crew:	5	Range:	1394 nm
Top speed:	372 mph	Service ceiling:	30,800 ft
Cruise:	310 mph		
Sea-level climb:	2515 ft per min		
Engine:	Allison T-56A-425 turboprop		

Miscellaneous information:
Normal crew: Pilot, co-pilot, Combat Information Center Officer, Air Control Officer, radar operator.

SERIAL INFORMATION

USCG BuNo	ex US Navy BuNo	Acquired
3501	160698	Jan 1987
3502	159497	Jan 1987
3503	159112	May 1988
3504	160011	Jul 1989
3505	159502	Sep 1989
3506	161342	Oct 1989
3507	158641	Oct 1989
3508	160415	Dec 1989

One E-2C '3501' was lost on 14 August 1990 when it exploded during its landing sequence at Roosevelt Roads air base, Puerto Rico. Crew of five were unfortunately killed.

During July 1989 the USCG E-2C unit moved from Norfolk, Virginia, to USCG Air Station, St Augustine, Florida.

HALL PH-2/3

The Hall Aluminium company produced an XPH-1 prototype flying-boat design in December 1929, on a US Navy contract, subsequently receiving orders for nine PH-1 boats from the prototype. After a lapse the Hall flying-boat went back into production during June 1936 for a contract from the Coast Guard for seven PH-2s to be used on air/sea rescue duties. They had more powerful Wright engines plus special equipment to fit the Coast Guard role, being otherwise similar to the PH-1. These remained in service until 1941, being survived by seven PH-3s ordered in 1939 by the Coast Guard. These had the same engines as the PH-2, but were fitted with long-chord NACA cowlings similar to those of the prototype XPH-1, and had a more refined cockpit enclosure.

These Hall flying-boats were the largest ever built for the service, acceptance trials taking place with the prototype at the Coast Guard air station at Cape May, New Jersey. The production model had been designed especially for patrol and rescue work in extremely rough weather, the type of weather the Coast Guard is often called out into fulfil its task. After using equipment that was six and even seven years old, the new Hall was quoted as 'badly needed equipment'. The new flying-boat had a quick take-off and a slow landing speed which combined to make operations in rough weather and choppy seas possible.

A regular air patrol along the Eastern seaboard was now possible, the patrol radius being extended from the regular 400 miles to 750 miles. The top cruising range was 2000 miles. The radio equipment installed was ultra-modern and included a two-way radiophone system, a radio direction finder, plus an amplifying system. The amplifiers were capable of transmitting the pilot's voice one mile down-wind and could be used to warn fishing fleets of approaching storms or to pass vital instructions to ships in distress.

The Hall Aluminium PH-2 boats were biplanes with a fabric-covered aluminium alloy structured wings and tail surfaces. The hull and wing tip floats were aluminium alloy. A special beaching gear consisted of two sets of double wheels and one tail wheel. Ball and socket joints were used to hold the gear in place enabling the flying boat to be handled on land. The boat could taxi into the water with the gear in place, being removed by a Coast Guard ground crew in a very short time. The hull design was very clean and quite unique. Of monocoque construction, it was built up around five bulkheads which divided it into six watertight compartments. Two of these were set aside with two bunks in each with accommodation for passengers or survivors who could be lifted on board in Army stretchers through the hatches and moved to the compartments on a track through the bulkheads. As many as twenty passengers/survivors could be accommodated when involved in rescue work, or assisting a shipwreck. The crew of each Hall PH-2 included pilot, navigator, radio operator and flight mechanic.

The power plants consisted of two supercharged Wright Cyclone engines of 750 hp each. Their maximum efficiency with the superchargers was at 2800 ft. At this altitude it was possible to fly for twenty hours at an airspeed of 152 mph. The propellers were Curtiss controllables, with Breeze cartridge-type engine starters. The landing speed was 60 mph.

Aircraft instruments included Sperry gyro-horizons, directional gyros, rate of climb indicators, exhaust gas analisers, and the latest in engine instruments and equipment. The electrical equipment was compact and complete, being carefully shielded and guarded against possible leaks and water spray. An intercom telephone system, fire extinguishers, life-saving and first-aid equipment, plus the normal Coast Guard armament, completed the list of standard equipment carried by each flying-boat. A novel feature was a folding anchor, this requiring only a few moments to launch on receipt of an emergency call.

Seven of the large PH-2 flying-boats served the USCG between 1934 and 1944. Photo depicts 'V169' flying over the snow capped Olympic mountain range in Washington state whilst based at Port Angeles Air Station prior to World War II. During the war years it was based at Miami, Florida. *(Gordon S Williams)*

The gasoline and oil systems were contained in the bottom wings and engine nacelles. The oil cooler occupied a strategic place in each engine nacelle, with a small scoop projecting out into the airstream. Surface controls in the tail were all balanced, the rudder being assisted by a small area built at the top and ahead of the hinge line. The elevators were balanced by a paddle balance above and ahead of the hinge line of each elevator. The tail group was large and well up clear of spray, this providing for excellent control of the boat in choppy seas. The ailerons were of the Friese type. All surface controls were cable-operated and the control wheels attached to an inverted U-shaped bridge which left the cockpit unobstructed. Six flying wires and two landing wires in each wing panel were used to take care of the enormous load factors encountered in rough weather operation.

After Pearl Harbor, the seven PH-3s were given standard US Navy finish in place of the natural metal. Some were adopted briefly for anti-submarine patrols. When the Hall PH-3s were eventually withdrawn from service, the era of the biplane flying-boat ended. To some students of aviation history the acquisition of the Hall PH-2/3 came as a surprise. The long years of service given by the twin-engined boats of the Douglas Dolphin and General Aviation types gave the mistaken impression that the most rugged type possible to design would likewise be another monoplane of full cantilever construction.

The Hall-Aluminium came from a rugged line of service aircraft, tracing their ancestry back as far as the Curtiss biplane of 1914. They had rendered good service to the US Navy, until made obsolete by the introduction of the larger monoplane boats which were in service in time for operations during World War II. With the Coast Guard the PH-2/3 completed its trial under fire which has been just as long as the Coast Guard duty performed by its predecessors. It was estimated that the wing loading of the new Hall-Aluminium was less than that of the older Coast Guard service.

The Coast Guard aircraft assignment listing dated 28 February 1943 gives the location of the following Hall-Aluminium flying boats. Elizabeth City, North Carolina, operated PH-2 V167 and PH-3 V183; Miami, Florida, operated a single PH-2 V164; Brooklyn, New York, operated PH-3 V177 and V174; Biloxi, Mississippi, operated two PH-2s V166 and V170; San Diego, California, operated PH-3 V178 and San Francisco operated two PH-3s V180 and V181.

OPERATIONS

At 8am on 15 July 1939 a radio message from the ketch *Atlantis* was transmitted to the Marine Hospital on Staten Island, New York. The *Atlantis* was the floating laboratory used by the US Oceanographic Institution at Woods Hole, Massachusetts. For

Hall PH-3 seen parked on the ramp at San Francisco USCG Air Station on 6 March 1942, finished in wartime drab camouflage. It was delivered to the service during January 1941 and the type served the USCG until 1944. This large flying-boat was purposely built for the service, acceptance trials taking place at Cape May, New Jersey. *(William T Larkins)*

several months her crew had been measuring area and currents of the Gulf Stream. Now one of them was seriously ill with pneumonia. The ketch was 150 miles south-east of New York, and at full speed it would take her eighteen hours to reach port. Instructions were returned by radio from the hospital for immediate treatment emphasising that the patient must be put ashore as soon as possible.

Just over two hours later at 1005am a US Coast Guard Hall PH-2 V-164 took off from the Floyd Bennett air station with a crew of seven to pick up the sick sailor. Shortly before noon the pilot William L Clemmer reported he had the ketch in sight and was preparing to land. The weather was dangerous with thunder squalls and a cross swell. But the Hall PH-2 landed safely and quickly took on board the ailing seaman who had been rowed out from the *Atlantis*. Then, taxiing over the swells into wind, the flying ambulance rose slightly, levelled off, and very mysteriously, with engines at full throttle, dived headlong into the sea. Both the USCG pilots and the sick man drowned. Five others were rescued by the *Atlantis*.

Two days later an official board of inquiry began piecing together the narrative fragments of the accident. It was deduced that V-164 had struck a long swell on the moment of getting airborne, had lost speed and nosed into the waves, its throttles still open. The most provocative testimony offered was that of Lieutenant Commander Watson A Burton, Commander of the USCG air station at Floyd Bennet, who deplored the unnecessary hazards to aircraft and men involved in the services famed aerial-ambulance service. 'In the majority of cases' he declared, 'it is discovered later that the patient is not as sick as reported. Also there are many cases that are sick as reported who would survive in spite of delay . . . Many times the information as to the ailment comes from the master of the ship who is not qualified to make a sound analysis.'

TECHNICAL DATA

Manufacturer:	Hall	Type:	Patrol and air-sea resuce flying boat
Contract No.:	Tcg. 26191 & 26810 PH-2 29347 PH-3	Unit cost:	PH-2 $116,104 PH-3 $170,000
Span:	72 ft 10 in		
Height:	19 ft 10 in	Length:	51 ft
Fuel:	892 galls	Wing area:	1170 sq ft
Empty weight:	9275 lb	Oil:	60 galls
Crew:	5	Gross weight:	16,457 lb
Top speed:	155 mph at 3200 ft	Stall speed:	61 mph
Cruise:	137 mph at 3200 ft	Range:	2242 miles
Sea-level climb:	1437 ft per min	Service ceiling:	21,000 ft
Engine:	Wright Cyclone F51 R-1820	Armanent:	Four .30 in Lewis machine guns
Take-off power:	2 × 875 hp	Three-blade prop blades:	Curtiss CS32D, constant speed
Prop diameter:	11 ft 9 in		
Designation:	PH-2/3		

SERIAL INFORMATION

USCG No.	Commissioned	Decommissioned
V164 PH-2	Apr 1938	15 Jul 1939 sank during rescue operation
V165 PH-2	Apr 1938	
V166 PH-2	Apr 1938	
V167 PH-2	Apr 1938	
V168 PH-2	Apr 1938	
V169 PH-2	Apr 1938	
V170 PH-2	Aug 1938	Jan 1944 Retired
V177 PH-3	May 1940	Mar 1944
V178 PH-3	Jan 1941	
V179 PH-3	Jan 1941	
V180 PH-3	Jan 1941	
V181 PH-3	Jan 1941	
V182 PH-3	Jan 1941	Nov 1941 Crashed
V183 PH-3	Jan 1941	

HOWARD GH-2/3 NIGHTINGALE

Based on the pre-war Howard DGA 'Damn Good Airplane' the GH- and NH- series of high-wing monoplanes were produced in quantity for the US Navy from 1941 onwards. Minor differences only distinguished the three utility models, each powered by a Pratt & Whitney R-985 engine. Production included 34 GH-1s, 131 GH-2s, and 115 GH-3s. The US Navy also used 205 NH-1 instrument trainers.

Twenty civil Howard DGA-15 models were impressed into US Army Air Corps service in 1942. These became designated as eleven UC-70 five-seat DGA-15P with R-985-33 engines, two UC-70A four-seat Model DGA-12 with R-915-1 engines, four UC-70B five-seat DGA-15J with R-915-1 engines, one UC-70C a four-seat model DGA-8 with a R-760-1 engine, and two UC-70D four-seat Model DGA-9 with R-830-1 engines.

The DGA-15 five-seat monoplane introduced by the Howard Aircraft Corporation in 1939 stemmed from the Bendix Trophy winner, the *Mr Mulligan* of 1934. Two of the descendants of this aircraft were acquired by the US Coast Guard in mid-1944 and were in use at the San Fransisco air station in California as proficiency and instrument training aircraft. They were designated GH-3. A year later in 1945, a single Howard GH-2 was delivered to the Coast Guard.

Very little information is available on this rare type used by the Coast Guard, and it is thought they were only retained in service for just over a year.

TECHNICAL DATA

Manufacturer:	Howard	Take-off power:	1 × 450 hp
Other designations:	UC-70	Designation:	GH-1/2 Nightingale
Span:	38 ft	Type:	Personnel transport
Height:	9 ft 5 in	Length:	25 ft 8 in
Empty weight:	2650 lb	Wing area:	185 sq ft
Top speed:	175 mph at 6000 ft	Gross weight:	4350 lb
Cruise:	154 mph at 15,000 ft	Range:	984 miles
Engine:	Pratt & Whitney Wasp Jr R-985	Service ceiling:	20,000 ft

Howard GH-3 BuNo 44952 of the US Navy. Known as the Nightingale, three served with the US Coast Guard as proficiency and instrument training aircraft at the San Francisco Air Station, California. They comprised two GH-3 and one GH-2 aircraft. *(Peter M Bowers)*

IAI COMMODORE WESTWIND 1123

During 1967 Israeli Aircraft Industries (IAI) based at Lod Airport, Israel, acquired all production and marketing rights for the Rockwell Standard Corporation Jet Commander (formerly Aero Commander) executive jet transport. The version developed for production in Israel was known as the Commodore Jet 1123.

This was the type, which, along with the Cessna Citation, was a recommended candidate leased for a six month period from 1 April to 30 September 1973 to determine if the operational envelope and characteristics were suited to the US Coast Guard MRS aircraft mission requirement. The Grumman HU-16E Albatross, affectionately known by its pilots as the 'Goat' and highly respected, was nearing the limits of its long life-span.

The first of two prototype Jet Commanders N601J had flown for the first time in the United States on 27 January 1963. The US production line was phased out between 1967 and mid-1969. By 31 May 1971 more than 140 aircraft, mostly produced in the USA, had been delivered. Sales by the new Israeli Aircraft Industries model had reached forty three.

Many modifications and improvements were incorporated by the new company into the original Commodore Jet 1121, these including an increase in the maximum take-off weight, strengthened undercarriage, greater fuel capacity and an improved performance. By June 1971 IAI commenced production of its new Commodore Jet 1123 Westwind which had a larger cabin, wingtip auxiliary fuel tanks, more powerful engines, an APU (auxiliary power unit), strengthened under-carriage,

Israeli Commodore Westwind 1123 'CG-160' was the second aircraft involved in a six month evaluation in search for a medium range search (MRS) aircraft to replace the Albatross. The evaluation team was based at Mobile, Alabama. Photo shows the Westwind parked at Juneau Airport, Alaska, during a stop-over in 1973. *(USCG)*

simplified electrical system, double-slotted flaps, drooped wing leading-edges plus two additional cabin windows. The prototype flew during September 1970, deliveries commencing a year later.

Commodore Jet Sales of Washington, DC, undertook the distribution of the first ninety-six production aircraft, servicing being provided by Commodore Aviation Corporation of New York. The 1123 Westwind endured a detailed operational and technical evaluation with the US Coast Guard, being finished in USCG livery and carrying the serial 619 on the tail. The Westwind was home-based at the USCG air station located at Mobile, Alabama. Along with the Cessna Citation CG 160, it took part in deployments throughout the Coast Guard Districts in the United States including Alaska. This was to introduce the aircraft to and evaluate it under different climes and operating conditions.

The MRS Task Organisation had Captain P A Hogue as its programme director, Commander B Harrington was the Project Manager, while Commander H J Harris Jr did the flight evaluation, assisted by Lieutenant Commander W C Donnell who did the engineering evaluation. Support activity at the home base of Mobile was organised by Captain A J Soreny, with operations handled by Commander J C Arney and engineering by Lieutenant G D Sickafoose.

Neither IAI Westwind or the Cessna Citation were selected for the Coast Guard MRS role, and after an abortive attempt to fill its requirement by purchasing Rockwell Sabre 75A aircraft through the US Navy, the Coast Guard issued a RFP (requirement for purchase) to the US industry in January 1975. From a number of submissions, the choice was eventually broken down to a version of the Dassault-Breguet Mystere 20 business jet, marketed in the United States by the Falcon Jet Corporation and known as the Falcon 20.

The Coast Guard selected the Falcon 20G variant, designated HU-25A Guardian by the Department of Defense, although this was unique in that it was not selected by any of the other US military services. A contract for forty-one aircraft was confirmed on 5 January 1977.

TECHNICAL DATA

Manufacturer:	Israeli Aircraft Industries	Designation:	Commodore Jet 1123 Westwind
Span:	44 ft 10 in	Type:	Medium range surveillance
Height:	15 ft 10 in		
Fuel:	1330 galls	Length:	52 ft 3 in
Empty weight:	11,070 lb	Wing area:	303 sq ft
Crew:	2	Gross weight:	20,500 lb
Cruise:	541 mph at 19,500 ft	Passengers:	10
Sea-level climb:	4100 ft per min	Stall speed:	86 mph
Engine:	General Electric CJ610-9 turbojet	Range:	1600 miles
		Service ceiling:	45,000 ft
Take-off power:	2 × 3100 lb thrust		

KAMAN K-225 MIXMASTER

Possibly one of the strangest craft ever used by the US Coast Guard was delivered to the Elizabeth City air station on 18 March 1950. This was a single Kaman K-225 helicopter manufactured in the factory located at Bradley International Airport, Windsor Locks, Connecticut, with date of completion 9 March 1950. It was one of three K-225s produced under a US Navy Bureau of Aeronautics contract for evaluation of its flying qualities. Charles H Kaman, later to become President and Chairman of the huge Kaman Aerospace Corporation, was himself an inventor and during 1945 had successfully developed a servo-flap control system for helicopter rotors.

This Coast Guard helicopter was the tenth produced by the company, the seventh K-225, and in addition to a BuNo quoted as 519-17 was allocated the USCG serial 239, later 1239. Only nine K-225s were built, of which two went to the US Navy, one to Turkey, the remainder staying in the United States. One went to Mississippi on geological survey work and two others were used for crop dusting.

On 11 May 1950, after evaluation at Elizabeth City by a number of USCG pilots, 1239 went into the workshops for an extended maintenance overhaul and by then had completed 120 flying hours. According to a USCG service report, after its initial evaluation, the Kaman was little used, and then chiefly for pilot refresher training. On 22 March 1954 the helicopter was removed from Coast Guard flight status and apparently returned to the US Navy, who operated it until 30 January 1955. By this time its total flight time was 165 hours.

It subsequently became surplus to requirements and was sold to the Charlotte Aircraft Corporation, Charlotte, North Carolina. An airworthiness certificate was issued to that company on 24 May 1955 together with Civil Aeronautics registration N1573M. It appears that the Charlotte Aircraft Corporation flew it an additional sixty hours until the first day of October 1955, when it crashed on a tidal flat near Fort Fisher, North Carolina, while transporting a fishing party to an island nearby. The pilot and two passengers were uninjured but the Kaman K-225 was totally demolished. Total flying hours were 237.40.

Apparently the K-225 was remarkable for its stability, with torque forces cancelled out by its contra-rotating rotors. It could be flown comfortably hands off, including in the hover. It was a three-seat open framework rotary wing aircraft with a tricycle landing gear and twin contra-rotating, intermeshing rotors, full servo-flap control surfaces mounted at seventy-five percent of rotor radius and control linkage by push-pull rods through a small gimbal ring assembly below the transmission and through the rotor blades to the flaps. Its overall length was thirty-eight feet, its heigh fourteen feet two inches, rotor diameter thirty-eight feet and the maximum gross weight 2500 pounds. It had a cruising speed of 60mph at 3000 engine rpm, with a top speed of 72mph. Other features included a fuel system thirty-eight gallons, gravity fed; electrical single wire system, direct current 12-volt system with an engine-driven generator and voltage regulator. It was powered by a Lycoming 0-435-A2 six-cylinder, horizontally-opposed air cooled engine of 450hp at 3000 rpm. The rotor operating range was 200-250rpm. Its range was 145 miles at cruising speed and the unit cost to the US Coast Guard was $37,684 and it was also known as the Mixmaster. It also carried the designation HK-1.

Seen in the hover at USCG Air Station Elizabeth City, North Carolina, on 30 January 1951 is the one and only Kaman K-225 Mixmaster CG-239 which was evaluated by the service. Parked in the background is a USCG Piasecki HRP-1 Flying Banana '1821' which served in small numbers. *(USCG)*

TECHNICAL DATA

Manufacturer:	Kaman	Engine:	Lycoming 0-435-A2
Other designations:	HK-1	Take-off power:	1 × 450 hp at 3000 rpm
Contract No.:	NOa(S)-10876	Designation:	K-225 Mixmaster
Rotor diameter:	38 ft	Type:	Three-seat utility helicopter
Height:	14 ft 2 in		
Fuel:	38 galls	Unit cost:	$37,684:90
Empty weight:	2200 lb	Length:	38 ft
Crew:	2	Oil:	3 galls
Top speed:	72 mph	Gross weight:	2500 lb
Cruise:	60 mph at 3000 engine rpm	Passengers:	1
		Range:	145 miles at 1500 ft
Sea-level climb:	650 ft per min	Service ceiling:	6150 ft

LOCKHEED XR30-1 ELECTRA

The Coast Guard acquired a total of six aircraft during 1936, which included five Viking OO-1 single-engine, single-float biplanes, the first of which had been previously acquired in 1931, and a single Lockheed Model 10-B Electra twelve-place passenger aircraft. It was commissioned during March 1936, serving as the US Coast Guard Commandant's flagship. It was designated XR30-1. A similar aircraft was acquired by the US Navy and designated XR20-1, being delivered on 19 February 1936, and served as a personal transport for the Secretary of the Navy. This Lockheed transport was equivalent to the commercial Model 10-A Electra.

Commissioned in March 1936, the Lockheed XR30-1 Electra '383' is seen at the Lockheed factory prior to delivery. It later became 'V151' and was used as the Commandant's flagship and re-designated R30-1. It was obtained by a trade-in with a Grumman JF-2 Duck to the US Marine Corps. The civil model was the Lockheed 10-B. *(William T Larkins)*

Lockheed XR30-1 was initially registered 383, later V151, being delivered to the Coast Guard on 19 April 1936, and in addition to being available to the Commandant was also personal transport for the Secretary to the Treasury. It was powered by two Pratt & Whitney Wasp Jr engines, and had a cruise speed of 170 knots.

The date of decommissioning is not known, but during 1939 it was with the Air Patrol Detachment at Cape May, New Jersey. It is listed as being assigned to the air station at Biloxi, Mississippi, on the USCG aircraft inventory dated 28 February 1943.

On 26 November 1935 the US Coast Guard delivered a brand new Grumman JF-2 Duck 175c/n 268 from the Grumman factory at New York, to the US Marine Corps base at Quantico, Virginia. Pilot was Lieutenant H D Palmer USCG. This was a trade for the Lockheed R30-1 383, later V151.

TECHNICAL DATA

Manufacturer:	Lockheed	Prop diameter:	8 ft 9 in
Other designations:	10A	Designation:	R30-1 Electra
Contract No.:	US Navy 44920	Type:	Transport
Span:	55 ft	Unit cost:	$65,000
Height:	10 ft 1 in	Length:	38 ft 7 in
Fuel:	200 galls	Wing area:	458 sq ft
Empty weight:	6203 lb	Oil:	14 galls
Top speed:	210 mph	Gross weight:	9750 lb
Cruise:	195 mph	Stall speed :	63 mph
Sea-level climb:	1140 ft per min	Range:	850 miles
Engine:	Pratt & Whitney Wasp R-985-48	Service ceiling:	21,450 ft
Take-off power:	2 × 400 hp	Two-blade prop blades:	Hamilton Standard controllable pitch

LOCKHEED R50-1/4/5 LODESTAR

Between 1940 and 1943 nearly one hundred examples of the Lockheed Model 18 Lodestar served with the US Navy, the US Marine Corps and the US Coast Guard. The series commenced with a single XR50-1 plus two R50-2s as command transports for the US Navy, and one, registered V188, for the US Coast Guard, this having Lockheed c/n 2008. It was delivered on 14 May 1940. There was a single R50-2 and three R50-3 transports which went to the US Navy.

The above plus twelve R50-4s were primarily executive transports with four to seven passenger seats in a VIP configuration. Four were delivered to the Coast Guard during 1942. The R50-5 was the standard twelve or fourteen-seat transport and three of these were acquired by the Coast Guard late in 1942. All three models were powered by Wright R-1820 engines.

Used mainly for administrative flights the Coast Guard Lockheed transports were based at Floyd Bennett Field, Brooklyn, New York, and at Elizabeth City, North Carolina, and later at Washington National Airport when the air station became established. The USCG aircraft inventory dated 28 February 1943 indicates that R50-1 V188 and R50-4 BuNo 12453 were based at Brooklyn, New York.

All the Lockheed transports were phased out between 1946

Lockheed R50-4 BuNo 12453 with 'Coast Guard' in small letters above the tail serial seen parked at Washington National airport during 1946. The four-stars on the engine cowl indicate it was the Commandant's flagship. It was commissioned in 1942 and served in Brooklyn, New York air station during 1943. *(William T Larkins)*

and 1953, being turned over to the US Navy, the War Assets Administration or the National Advisory Committee for Aeronautics — NACA. One of these transports crashed during 1948, and the R50-1 V188 was sold to South Africa becoming registered ZS-BAJ.

TECHNICAL DATA

Manufacturer:	Lockheed	Prop diameter:	10 ft 7 in
Other designations:	Model 18	Designation:	R50-1/4/5 Lodestar
Span:	65 ft 6 in	Type:	Transport
Height:	11 ft 11 in	Unit cost:	$185,000 R50-1
Fuel:	644 galls	Length:	49 ft 10 in
Empty weight:	11,821 lb	Wing area:	551 sq ft
Crew:	17	Oil:	44 galls
Top speed:	246 mph at 7900 ft	Gross weight:	17,500 lb
Cruise:	231 mph at 12,000 ft	Stall speed:	69 mph
Sea-level climb:	1600 ft per min	Range:	2000 miles
Engine:	Wright Cyclone R-1820-40	Service ceiling:	25,400 ft
Take-off power:	2 × 1100 hp	Three-blade prop blades:	Hamilton Standard, hydromatic

SERIAL INFORMATION

USCG No		Commissioned	Decommissioned
V188	R50-1	14 May 1940	Brooklyn, N.Y. 28 Feb 1943. Sold as ZS-BAJ
12453	R50-4	1942	Brooklyn, N.Y. 28 Feb 1943.
05049	R50-4	1942	
12447	R50-4	1942	
	R50-4	1942	
12456	R50-5	1942	
	R50-5	1942	
	R50-5	1942	

LOCKHEED HC-130B/E/H HERCULES

The Lockheed C-130 Hercules plays a major role in US Coast Guard Aviation. As part of its fleet of over 200 fixed-wing aircraft and helicopters, the Coast Guard operates no less than thirty-one out of forty-two the service has acquired over the years. Six USCG air stations are equipped with the 'Herky Bird' employed on a wide envelope of tasks, their daily accomplishment alone showing the tremendous impact of the Coast Guard on American lives. They each average approximately 1000 hours flying annually.

HISTORY

It was during 1951 that development of the Lockheed Model 82 commenced, after a top US Air Force policy decision to re-equip with turboprop transports, a decision which produced the Lockheed C-130, the projected Douglas C-132 which was cancelled, and the Douglas C-133 Cargomaster. The Hercules was the first transport produced with the weapon system concept, as the SS-400L medium cargo support system.

The first of forty-eight Lockheed C-130 Hercules utilised over the years as a workhorse by the USCG was delivered on the last day of 1959. Depicted is '1339' based at Elizabeth City, North Carolina, seen on 15 August 1961. It served until April 1986, when retired going to the US Air Force. *(USCG)*

A contract for two prototypes designated YC-130 was placed on 11 July 1952 and this was followed by production contracts in September 1952. The two YC-130s, 53/3396/7, were built at Burbank, California, where Stanley Beltz made the first flight on 23 August 1954. The production line was laid down at Marietta,

Georgia, where the first C-130A 53-3129 first flew on 7 April 1955. Like the two prototype YC-130s it was powered by four 3750shp Allison T56-A-1A turboprops and had a bluff front fuselage. Later 'A' models had T56-A-9 engines.

US Navy interest in the Lockheed Hercules crystalised during 1960 when the US Air Force demonstrated the potential of the turboprop transport in both Arctic and Antarctic conditions, operating as a ski-equipped transport. One US Air Force squadron was operating ski-equipped 'D' models, primarily in support of DEW line construction sites in the Arctic, and five of these aircraft were despatched to the Antarctic to assist US Navy in Operation Deep Freeze 60. This contribution was so outstanding that the US Navy went ahead and purchased four ski-equipped C-130B transports for use by VXE-6 Squadron for Operation Deep Freeze 61. These were initially designated UV-1L.

Independent of the US Navy interest in the Lockheed C-130 Hercules, the US Marine Corps became aware of the Hercules potential as a flight refuelling tanker during 1957, and two US Air Force 'A' models were modified for evaluation. This led to a US Marine Corps order for forty-six tankers, designated GV-1.

The US Navy purchased seven Hercules transports, similar to the USMC GV-1 tanker, but with the refuelling equipment removed, and designated GV-1U. A more specialised role for which the US Navy selected the Hercules was that of airborne communications relay aircraft to support the worldwide operations of the Fleet Ballistic Missiles submarines.

One of the largest operators of the surveillance-patrol version of the ubiquitous Lockheed Hercules transport is the US Coast Guard. It was during 1957 that the choice of the C-130 Hercules

Only four Hercules in the '1400' series were used by the USCG including '1452' depicted which was delivered in March 1968, and seen whilst serving at San Francisco Air Station. This was a HC-130H similar to the USAF rescue version, but minus the Fulton rescue yoke in the nose. *(AP Photo Library)*

to equip the Coast Guard was first confirmed, the basis of the selection being the aircraft's long-range capability at low altitude, its mission endurance and its ability to conserve fuel and increase time-on station by shutting down two engines. Early US Coast Guard evaluation trials confirmed and demonstrated the C-130's excellent low-speed handling qualities when required to fly low and slow over target areas. It had a mission endurance of eighteen hours and more than 2600 nautical miles range.

LOCKHEED R8V-1G

During 1958 four Lockheed Model 282-2B aircraft were ordered by the US Coast Guard under the US Navy designation R8V-1G, these aircraft being allocated the USCG numbers 1339 to 1342, and delivered for service late 1959/early 1960 at the air stations located at Elizabeth City, North Carolina; Barber's Point, Hawaii; San Francisco, California; and St Petersburg, Florida, respectively. In due course these transports were redesignated SC-130B as being derivatives of the C-130B powered by T56A-7 engines. Two identical aircraft were ordered in 1960 as 1344 and 1345 and delivered in 1961, both initially going to Elizabeth City air station. Under a US Department of Defense directive dated 6 July 1962, all designations allocated to US Navy aircraft and variants were standardised with those of the many US Air Force variants, resulting in these six Herclues being redesignated HC-130B.

In March and April 1962, three more aircraft were delivered to the US Coast Guard — 1346/7/8, the first being assigned to San Francisco, while the other two went to the detachment based at Argentia, Newfoundland, for work in the North Atlantic, including the International Ice Patrol. Later 1347 moved to Elizabeth City air station in 1979 where it was equipped with sideways-looking radar (SLAR) fitted on the undercarriage fairings and with sensor pods under the wings.

The last three HC-130B transports were delivered to the US Coast Guard during late 1962/early 1963 as 1349 to 1351. These completed the initial USCG procurements which came via US Air Force contracts and all designated Lockheed Model 282-2B. All were retired during April 1986, going to the huge storage depot located at Davis-Monthan AFB, Arizona.

A typical sortie pattern by a US Coast Guard HC-130B would be to fly out 1000 miles of 370mph at 25,000 feet, descending to 5000 feet cruise for eight hours search duty at 145mph on two engines, restart engines and return to base at 300mph, leaving a twenty-five minute fuel reserve. This type of flying involved operating in the most testing of all conditions. The Hercules are mainly based at US coastal stations, flying in turbulent air at low level as well as in constant salt air. Internal accommodation is normally provided for an extra radio operator and two search equipment operators, while it can carry between twenty-two and forty-four passengers. The rear doors, normally used by paratroopers and equipment dropping, incorporate large clear vision panels, useful on visual searches.

One special aircraft, replacing an equally special Douglas EC-54 Skymaster, was a Lockheed Model 382-4B designated EC-130E and numbered 1414 delivered to Elizabeth City for special equipment installation on 23 August 1966. This Coast Guard Hercules was employed to calibrate the long-range navigation (LORAN) equipment and flew worldwide on this task, being home based at Elizabeth City. It was retired during April 1986.

The last Hercules in the '1500' series was '1504' seen in a spectacular photo whilst on patrol with the International Ice Patrol on 30 August 1983. The SLAR — sideways looking infra-red radar — can be seen fitted to the port undercarriage bay. Delivered in April 1974 this aircraft is still in service. *(Lockheed)*

Shortly after the first C-130H had been delivered to the US Air Force Aerospace Rescue & Recovery Service (ARRS), the US Navy gained an appropriation of funds on behalf of the Coast Guard for three of the HC-130H T56A-15 engined aircraft, and 1452 to 1454 were delivered during March and May 1968. The second of these 1453 was the 1000th Hercules produced. During 1973/74 five Model 383C/27D numbered 1500 to 1504, were ordered, these completing the second ten-year procurement of Hercules for the Coast Guard. Unlike the US Air Force ARRS Hercules, the US Coast Guard transports possessed no provision for the Fulton ground-air recovery yoke system.

No more Hercules were delivered to the Coast Guard until late 1977 when four Model 382C-70D HC-130Hs 1600 to 1603 were destined for service in Kodiak, Alaska. This was followed by six Model 382C-37E HC-130H-7 Hercules, 1700 to 1705, the first being delivered on the last day of May 1983, at Clearwater, Florida, replacing twenty-five year old HC-130B transports. However the new transports utilised the -7 engines from the older aircraft then in storage at Davis-Monthan.

Between 20 August 1984 and May 1988 the Coast Guard took delivery of no less than seventeen more HC-130H aircraft which included four Model 382C-50E 1706 to 1709, two Model 382C-57E 1710 and 1714, three Model 382C-61E 1711, 1712 and 1713, one Model 382C-64E 1715, one Model 382C-76E 1716, three Model 382C-79E 1717, 1718 and 1719 and two Model 382C-84 1720 and 1721. One Hercules was allocated an out of sequence Coast Guard serial, this being a Model 382C-22E 1790 coinciding with the year the US Coast Guard was formed. It was delivered to Kodiak, Alaska, on 23 July 1983.

Today the Coast Guard Hercules are identified in the USCG inventory in the 1500, 1600 and 1700 series by their serial numbers, the oldest in service 1502 being brought out of retirement and storage at the Davis-Monthan storage centre for rework and return into service.

OPERATIONS

Since commencing service with the US Coast Guard during late 1959, only one Lockheed C-130 has been lost. This was 1600 which crashed on 30 July 1982 at Attu in the Aleutian Islands, a USCG outpost known for its atrocious weather conditions. It crashed some four kilometres south of Attu while attempting a landing in bad weather.

Four Model 382C-70D HC-130H Hercules were delivered to the USCG late in 1977, with '1600' depicted being delivered on 26 October 1977. Lockheed marketed this model as the C-130H-MP designed for patrolling off-shore 'exclusive economic zones' (EEZ). Hercules '1600' crashed on 30 July 1982, whilst making an approach to Attu in the Pacific Aleutian chain. *(Lockheed)*

Today the Hercules operates out of six US Coast Guard air stations on a wide variety of tasks. The air stations include Elizabeth City, North Carolina, which has the huge USCG Aircraft Repair & Supply Center, an engineering facility where modifications and salt corrosion control is carried out. Five 'Herky Birds' are currently based here. Clearwater, which up to 1979 was St Petersburg air station in Florida is responsible for surveillance of the Gulf of Mexico and the Caribbean with its six HC-130H aircraft. McClellan AFB near Sacramento, California, is the home of four HC-130H-7 Hercules 1700, 1702, 1703 and 1704 under the command of Captain Kirk Colvin USCG. A forward-looking infra-red (FLIR) turrent for night operations has been fitted to these aircraft. It is a Lockheed modification and is used for patrolling fishing areas, law enforcement and other tasks. Prior to 1978 the USCG Hercules were based at San Francisco air station, but despite moving inland to McClellan, are still responsible for the Pacific seaboard of the United States.

Kodiak air station in Alaska is responsible for the northern Pacific, the Aleutian Islands, and operates a fleet of six Hercules. Barbers Point in Hawaii is responsible for the central Pacific area with three HC-130H aircraft, while the same number are now based at Borinquen air station in Puerto Rico. These 'Herky Birds' plus the ones at Clearwater, Florida, are all fitted with AN/APS-137 FLIR search radar, as carried by the US Navy Lockheed P-3C Orion patrol aircraft. The AN/APN-215 weather radar is maintained.

OIL, ICE & FISH

Four C-130H Hercules are used to fill gaps required by maintenance, modifications, etc. The last HC-130H to be delivered to the Coast Guard in May 1988, 1721, is being equipped with Grumman E-2C Hawkeye-type radar fitted to the top of the fuselage, the first ever Hercules to be so modified. This will enhance the airborne barrier in the drug war.

When the huge oil spill first occurred from the tanker *Exxon Valdez* in Alaska during March 1989, the US Coast Guard HC-130H aircraft from McClellan were immediately called in to take the National Strike Team, which is based at Hamilton AFB, San Francisco, along with its booms and equipment to begin the nigh impossible task cleaning up the huge oil spill. The effort continued for many months, and the US Coast Guard Hercules logged many hours flying supplies to Alaska to help the huge clean up operation which involved many USCG personnel.

The Coast Guard depends on its long-range HC-130H aircraft for its ice patrol reconnaissance. Each year, a HC-130H from Elizabeth City is deployed on ice patrol approximately every two weeks. A crew of eleven, along with three ice observers from the International Ice Patrol HQ, Groten, Connecticut, deploy with the aircraft. The task is to provide long-range reconnaissance over the Grand Banks region south-east of Newfoundland, Canada. The crews search the area both visually and with sideways-looking radar (SLAR). The radar equipment enables the C-130 to cover twenty-seven miles on each side, up to eighty miles with poorer resolution, of the aircraft's flight path, making it possible to cover vast areas of water in a single flight.

Normally operating at 8000ft the Hercules can fly six hours often never seeing the ocean, yet painting targets on its SLAR. Information coming via the SLAR is recorded on film giving the icebergs' position in latitude and longitude, the time speed and aproximate size. Later, the specialists can make an overlay of the photos and determine the direction and speed of selected icebergs. Some are marked with a chloride-rhodamine 'B' dye bomb, this involving flying low over the iceberg target.

The year 1983 will go down as the second biggest year for icebergs since the sinking of the liner *Titanic*, with almost 2000 icebergs sighted and recorded. During the week beginning 11 July 1983 with only a day and a half of weather clear enough for visual sighting during six days of flying, a Ice Patrol HC-130H crew chartered seventy-six icebergs ranging from twenty-foot 'growlers' to multi-million pound icebergs the size of an aircraft carrier. The arrival of the high technology sensing equipment, built by Motorola and installed on either side of the Hercules fuselage, over the landing gear, came at an appropriate time.

Another technology advance has helped to turn the International Ice Patrol exercise into a real science — the computer. During the winter of 1978 'Project Ice Warn' took place involving side-looking radar antenna fitted under the tail of the Hercules HC-130B '1351', the project being financed by the International Ice Patrol with the NASA providing the engineering and scientific advisors for the project. This was the forerunner of the current SLAR.

Patrolling America's vast and valuable fishing zones is one of the vital missions of the Coast Guard, and has received even greater emphasis with the nation's incorporation of the 200-mile off-shore Exclusive Economic Zone — EEZ. The most important fisheries surveillance takes place off the state of Alaska. There, the US Coast Guard fleet of Hercules, along with USCG cutters and cutter-borne helicopters are involved. Eight and a half hour daily C-130 patrols are flown at low level on special tracks ranging from Kodiak to Attu on the western side of the Aleutian chain of islands. The 'Herky Bird' crews seek to identify each fishing boat, along with its position and activity. The shipping in this large bountiful fish area involves vessels from Japan, the Soviet Union and the USA.

The Coast Guard Hercules are also equipped with the Airborne Oil Surveillance System (AOSS) which is so sophisticated that the smallest portion of oil on the surface of the ocean can be easily detected night or day. How long, where the oil originated, etc is just some of the detection data available. This is part of the Marine Environmental Protection (MEP) mission focused on the National Strike Force, a select sixty-man USCG group of personnel that is specially trained for fight oil spills. The Pacific team is based at Hamilton AFB, California, and the Atlantic team at Mobile, Alabama.

Since the creation of the strike force in 1973, the men and the Hercules have flown to hundreds of spills, taking along their sophisticated equipment. In 1975, at the request of the Japanese government, a US Coast Guard strike force with its Hercules flew to the Straits of Malacca, near Singapore. A huge supertanker *Showa Mora* had gone aground, spilling more than a million gallons of oil. The strike team helped to syphon off the remaining oil. The teams carry an Air Deliverable Anti Pollution

The last of six Model 382C-37E HC-130-7 is '1705' seen flying over the coastline of bleak Alaska from its base at Kodiak USCG Air Station. The -7 engines came from older USCG Herky Birds. Hercules '1705' was delivered on 29 June 1984 and today is one of a fleet of thirty-one in the inventory. *(Lockheed — John Rossino)*

Transfer System (ADAPTS) that can pump 1800 gallons of oil a minute. Another unit, the High Seas Oil Containing System (HSOCS) is packed like an accordian. When put into service, it unfolds onto the water like a snake and inflates to surround and keep the oil confined. Six hundred feet long, the boom can survive twenty knot winds and five foot seas.

Search and Rescue (SAR), law enforcement, drug interdiction, all involve the workhorse of Coast Guard Aviation, the Hercules. Not all routine patrols are exciting. The long reconnaissance patrols across the Caribbean involving the drug interdiction campaign entail looking for anything that moves. One USCG Lockheed HC-130H pilot described law-enforcement patrols as 'hours and hours of complete boredom interrupted by a box lunch.'

Flying a Lockheed C-130 in the US Coast Guard is not only a man's work, as the USCG pilot corps has at least three women 'Herky Bird' pilots, one based at Boringuen, one at Clearwater and one at Sacramento. Pilots in the US Coast Guard Reserve also fly the C-130. The US Coast Guard are getting a good performance factor out of the fleet of thirty-one Hercules, and Lockheed are proud that the aircraft is a major factor in the Coast Guard's ability to accomplish its many tasks. A reminder that the US Coast Guard motto is *Semper Paratus* — Always Ready.

TECHNICAL DATA

Manufacturer:	Lockheed	Designation:	HC-130H
Other designations:	Model 382	Type:	Maritime patrol, Search & Rescue
Span:	132 ft 7 in		
Height:	38 ft 3 in	Unit cost:	$6 million plus
Fuel:	9680 galls	Length:	97 ft 9 in
Empty weight:	76,780 lb	Wing area:	1734 sq ft
Crew:	4 + 3 = 7	Gross weight:	175,000 lb Six standard freight pallet
Top speed:	386 mph at 25,000 ft		
Cruise:	353 mph		
Sea-level climb:	2570 ft per min	Range:	2745 st miles with max payload
Engine:	Allison T56-A-15 turboprop	Service ceiling:	33,000 ft
Take-off power:	4 × 4508 eshp	Four-bladed prop blades:	Hamilton Standard Hydromatic
Prop diameter:	13 ft 6 in		

Miscellaneous information:

The Auxiliary Power Unit (APU) supplies air for engine starting, ground air conditioning and an auxiliary AC generator.

SERIAL INFORMATION

USCG No.	Commissioned	Decommissioned	
1339	31 Dec 1959	April 1986	to US Air Force
1340	31 Dec 1959	April 1986	to US Air Force
1341	31 Jan 1960	April 1986	
1342	24 Mar 1960	April 1986	to US Air Force
1344	25 Jan 1961	April 1986	to US Air Force
1345	9 Mar 1961	April 1986	to US Air Force
1346	23 Mar 1962	April 1986	to US Marine Corps
1347	1 Mar 1962	April 1986	to US Air Force
1348	19 Apr 1962	April 1986	to US Air Force
1349	14 Dec 1962	April 1986	to US Air Force
1350	31 Jan 1963	April 1986	to US Air Force
1351	22 Feb 1963	April 1986	
1414	23 Aug 1966	April 1986	
1452	Mar 1968	April 1986	to US Air Force
1453	1 May 1968	April 1986	to US Air Force
1454	May 1968	April 1986	to US Air Force
1500	Aug 1973		
1501	Sep 1973		
1502	Oct 1973		
1503	Mar 1974		
1504	Apr 1974		
1600	26 Oct 1977	Crashed 4km south of Attu 30 July 1982	
1601	1 Nov 1977		
1602	15 Nov 1977		
1603	7 Dec 1977		
1700	31 May 1983		
1701	1 Apr 1984		
1702	20 Jun 1984		
1703	29 Jun 1984		
1704	20 Jul 1984		
1705	29 Jun 1984		
1706	20 Aug 1984		
1707	18 Sep 1984		
1708	18 Oct 1984		
1709	10 Dec 1984		
1710	16 Aug 1985		
1711	16 Sep 1985		
1712	7 Oct 1985		
1713	16 Oct 1985		
1714	29 Oct 1985		
1715	15 Nov 1985		
1716	Dec 1986		
1717	3 Nov 1987		
1718	25 Nov 1987		
1719	9 Dec 1987		
1790	23 Jul 1983	Out of sequence serial; Year USCG formed	
1720	Apr 1988		
1721	May 1988	Converted with E-2C Hawkeye radar on top of the fuselage	

LOENING OL-5

The Loening OL-5 amphibian biplane has the distinction of being the very first type ordered specifically for the US Coast Guard. The US Congress appropriated $152,000 for the first purchase of aircraft for the Coast Guard during 1926. Three OL-5s and two Chance Vought UO-4s were purchased. Two of the OL-5s were assigned to the first Coast Guard air station on Ten Pound Island, Gloucester, Massachusetts, the other being deployed to the USCG air station at Cape May, New Jersey.

Grover C Loening, founder of the company which bore his name, made great contributions to aircraft design, and the distinctive amphibian was characteristic of his original work. This resulted from an attempt to match contemporary landplane performance with an amphibian of similar horsepower. Perhaps of the numerous designs produced by the Loening Aeronautical Engineering Company of New York between 1917 and 1928, the series of US Army and US Navy amphibians which initially appeared in 1923 became the best known.

A novel approach was adopted by Loening, using a large single float under the fuselage, rather than a flying-boat hull as adopted by other amphibians of the day. This float was faired into the fuselage with wheels located on each side. They were arranged so that they could be swung out of the way when operating on water. The first of the Loening amphibians were powered by inverted Liberty engines, and were used in an observation role by the US Army Air Corps and designated COA-1s. The US Navy purchased five identical aircraft in 1925 for use in connection with the 1925 Arctic Expedition; these were designated OL-2. The OL-1 had already been used by the US Navy on two Packard-powered Loening amphibians. The US Navy ordered four more which had improvements, these becoming the OL-3. A further six powered by the Liberty engine were designated OL-4.

One historical source indicates that no examples of the OL-5 were built, the OL-6 being a Packard-powered amphibian for the US Navy and the US Marine Corps, having a new, more angular tail design. One of these became the XOL-7 with experimental thick-section wings, while another became the XOL-8 fitted with the Pratt & Whitney Wasp, the first radial engine to be used in the Loening series. This engine powered the twenty production OL-8s which were two-seat amphibians, and the twenty OL-8As fitted with arrester gear for carrier deck operation. By the time the US Navy acquired twenty-six OL-9s outwardly similar to the OL-8, the Loening company had merged with the Keystone Aircraft Corporation at Bristol, Pennsylvania, soon to be acquired by the huge Curtiss-Wright Corporation. Finally the US Navy operated two other examples of the same basic Loening design for ambulance duties designated XHL-1. They had a single open cockpit for the pilot and carried six passengers in the fuselage.

OPERATIONS

The three new Coast Guard Loening OL-5 aircraft were allocated serial numbers CG1, CG2 and CG3 respectively. CG1 was delivered on 14 October 1926, serving at Ten Pound Island in Gloucester Harbor. CG2 was delivered sometime during October, and went to Cape May, New Jersey, and CG3 was delivered on 3 November 1926, also serving at Ten Pound Island. The official commissioning ceremony took place in December. By 1 September 1928, CG1 had flown 136 hours 50 minutes, CG2 453 hours 55 minutes and CG3 526 hours 30 minutes.

Acquisition of the OL-5s created a specific need for standard aircraft radio equipment but, because of the nature of their duties, such as scouting and rescue operations over water, neither US Navy nor available commercial types seemed suitable. As a result, the first standard Coast Guard radio equipment was designed and assembled by a Coast Guardsman, Radio Electrician A C Descoteaux. This radio, battery-operated to make it entirely independant of the aircraft electrical system and to insure its usefulness in the event of a forced landing at sea, provided two-way continuous wave telegraph and high quality voice communiction. The spoken word could be transmitted 150 miles, while signals from the morse key could reach as far as 1200 miles. The set consisted of a transmitter and receiver that weighed a total of ninety pounds.

These first aircraft purchased especially for the US Coast Guard aviation had been specially constructed, being of stronger build and having a greater fuel capacity then US Navy aircraft of the same make. The 400hp inverted Liberty engine consumed twenty gallons of gasolene each hour from tanks holding 135 gallons at cruising speed. In addition to the pilot and observer, the amphibian could carry one passenger.

The good work accomplished by the three OL-5s persuaded the Commandant of the Coast Guard, Rear Admiral F C Billard, that the air arm ought to be enlarged because of the manifold uses of aircraft. The duties listed by the Commandant were possibly the very first SOPs (Standing Operating Procedures) issued by USCG HQ. So the OL-5 crews spent time hunting schools of fish for local fishing fleets, and on one occasion during their law enforcement activity, machine-gunned and sunk 250 cases of liquor thrown overboard from a rum runner. They took aerial photographs and patrolled beaches along the coast line.

Flying an OL-5 out of Cape May in the spring of 1929, Ensign W E Anderson spotted the yacht *Nomad* wallowing in the trough of a heavy sea. He radioed a patrol boat which came alongside the sinking vessel. A boarding party found all hands inebriated, one dead and the helm awash. *Nomad* would certainly have foundered except for Coast Guard aviation.

The Loening OL-5 was one of the first aircraft type acquired by the US Coast Guard, three being commissioned during December 1926. Initially allocated the serials 'CG-1/2/3/', two crashed and the second aircraft was retired in April 1935 due to age. Depicted is a USCG OL-5 seen shortly after take-off. *(AP Photo Library)*

A short time later, a Loening OL-5 demonstrated how USCG amphibians could capture criminals. On 3 September 1929, Chief Gunner C T Thrun and his two crewmen were testing the radio in OL-5 CG2. While flying over Murder Kill Inlet, Delaware, he received a message that a sloop had been stolen from the Riverton Yacht Club, some way up the Delaware River, above Philadelphia. After receiving an accurate description of the stolen vessel, Gunner Thrun set out in pursuit. Within twenty minutes he had found the fleeing sloop off New Castle, Delaware, a full half day's sail from any Coast Guard surface vessel. The OL-5 landed alongside the stolen sloop *Bronco* and sent Chief Machinist's Mate C H Harris armed only with a Very pistol in the seaplane's rubber lifeboat with instructions to board and place the crew under arrest. The OL-5 took off in order to radio USCG patrol boats from Section Base Nine, and eventually the sloop and prisoners were handed over to *CG-182*.

The first OL-5 to perish was the CG3, which crashed on 10 November 1929. It was wrecked when it struck the mast of the fishing vessel *Jackie B* in Gloucester Harbor. Lieutenant Melka, Machinist Kenly and Radio Electrician Descoteaux were taken to hospital suffering from shock.

On 21 June 1930, OL-5 CG1 hit the superstructure of the yacht *Whiz* at new London, Connecticut, while patrolling a yacht race. The pilot Lieutenant Norman N Nelson and his passenger, Commander Eugene A Coffin, got clear in their safety belts and surfaced without serious harm. The remaining OL-5 CG2 was decommissioned in April 1935 because of its age.

TECHNICAL DATA

Manufacturer:	Loening	Type:	Observation amphibian
Span:	45 ft	Unit cost:	$32,710
Height:	12 ft 9 in	Length:	35 ft
Fuel:	135 galls	Wing area:	504 sq ft
Empty weight:	3805 lb	Oil:	12 galls
Crew:	2	Gross weight:	5471 lb
Top speed:	120 mph	Passengers:	1
Cruise:	65 mph	Stall speed:	59 mph
Engine:	Liberty V-1650-1	Range:	450 miles
Take-off power:	1 × 400 hp	Service ceiling:	12,750 ft
Designation:	OL-5		

SERIAL INFORMATION

USCG No.	Commissioned	Decommissioned	
CG1 101	14 Oct 1926	21 Jun 1930	Crashed
CG2 102	Oct 1926	Apr 1935	Retired
CG3 103	3 Nov 1926	10 Nov 1929	Crashed

MARTIN PBM-3/5 MARINER

The archives reveal that the US Coast Guard took delivery of twenty-seven Martin PBM-3 twin-engined amphibians during the first half of 1943. These had the designations PBM-3C, PBM-3D and PBM-3S. Late in 1944 the first batch of forty-one PBM-5 Mariners were delivered to the Coast Guard. The final Mariners from this latter batch were delivered during the latter part of 1945. Ten years later in 1955, there were still ten in service. The earlier PBM-3s had been returned to the US Navy during the first half of 1946.

By the mid-1930s a rivalry had developed between the Glenn L Martin Company of Baltimore, Maryland, and the Consolidated

A USCG Martin PBM-5 Mariner BuNo 84732 makes a jet-assisted take-off from the Air Station at Salem, Massachusetts. This assistance enables a shorter take-off run, and enables loads heavier than normal to be carried. The 'R-22' on the nose is not only the aircraft ident, but the radio call-sign 'RESCUE 22'. *(USCG)*

Aircraft Corporation of San Diego, California. Well established as a builder of US Navy flying-boats Glenn L Martin was involved with replacements for earlier US Navy types. Designed in 1937, his Model 162 challenged the Consolidated PBY Catalina, demonstrating a performance which was superior.

The new flying-boat featured a deep hull with a gull wing and was powered by two Wright Cyclone engines. To test the handling qualities of the new design, the company built a single-seat, quarter-scale model known as the Model 162A. Today this is preserved in storage with the National Air & Space Museum at Silver Hill. On 30 June 1937 the US Navy placed a contract for one full-size prototype, designated XPBM-1. First flight was on 18 February 1939, powered by 1600hp R-2000-6 engines with gun positions in nose and dorsal turrets plus in the waist and tail. It was designed to carry 2000lb of bombs or depth charges. Initially the new flying-boat had a flat tailplane with outrigged fins, but later dihedral was added, this giving the Martin Model 162 one of its most striking features.

By the end of 1937 the US Navy had ordered twenty production model PBM-1s for which the name Mariner was selected. All were completed in April 1941. There was a single XPBM-2 with extra fuel and provision for catapult launching. On 1 November 1940 orders were placed with Martin for 379 PBM-3s, these appearing from April 1942 onwards in several different versions. The retractable wing stabilising floats became fixed and strut-braced, asnd the airframe had lengthened engine nacelles providing stowage for bombs or depth charges. Power for the basic PBM-3 was from R-2600-12 engines, and the variant included fifty unarmed PBM-3R transports with seats for twenty passengers, 274 PBM-3Cs with standardised US/British equipment, and 201 PBM-3Ds with R-2600-22 engines with improved armament and armour protection. A large number of the PBM-3Cs and PBM-3Ds carried search radar in a large housing above and behind the cockpit, and the US Navy experience with the use of this radar led to the development in 1944 of a long-range anti-submarine version, the PBM-3S. A total of 156 of this latter variant were delivered, powered by R-2600-12 engines.

The PBM-4, of which 180 were contracted for during 1942, was not proceeded with. It would have had R-3350-8 engines. In 1943 the XPBM-5 appeared with a choice of 2100hp Pratt & Whitney R-2800-22 or -34 engines, and contracts for the production of the PBM-5 were placed with Martin in January 1944. It was delivered from August 1944 until the end of hostilities. This version was equipped with eight .50-in machine guns and AN/APS-15 radar. The latter equipped the PBM-5E version, while some PBM-5S variants were also in use, the designation indicating an anti-submarine role. Production totalled 631.

An amphibious version was the final Mariner development, designated PBM-5A, thirty-six of which had been built by the time production was completed in April 1949. This version served primarily in the air/sea rescue role with the US Navy and the US Coast Guard, replacing PBM-5G flying-boats with the USCG.

As of 31 October 1944 the Mariner was based at the following US Coast Guard air stations within the United States: Elizabeth City, North Carolina — five PBM-3s; Miami, Florida, had a single PBM-3C; Port Angles, Washington — two PBM-3S; St Petersburg, Florida — two PBM-3S; Salem, Massachusetts — two PBM-3S; San Diego, California — one PBM-3C and three PBM-3S; San Francisco, California, had two PBM-3S. Of the forty-one PBM-5 aircraft in use by the Coast Guard a total of fifteen were lost. The US Coast Guard aircraft inventory for 1 May 1947, lists twenty-four PBM-5 Mariners in service. This had been reduced to fifteen PBM-5G by January 1950. One of these latter was seen at San Francisco air station on 3 April 1947 with the unusual BuNo V84728.

Martin PBM-5G BuNo 84736 seen moored on the slipway at San Francisco USCG Air Station on 10 October 1946. Forty-one of this Mariner variant served with the USCG serving until 1956. The first Mariner PBM-3S entered Coast Guard service during early 1943. *(William T Larkins)*

TECHNICAL DATA

Manufacturer:	Martin	Length:	80 ft
Span:	118 ft	Wing area:	1408 sq ft
Height:	27 ft 6 in	Gross weight:	58,000 lb
Empty weight:	32,378 lb	Range:	2137 st miles
Crew:	9	Service ceiling:	16,900 ft
Top speed:	198 mph at 13,000 ft	Armament:	Two flexible .50in guns in nose and dorsal turrets. Single .50in in waist and tail. 2000 lb bombs or depth-charges.
Sea-level climb:	410 ft per min		
Engine:	Wright R-2600-12		
Take-off power:	2 × 1700 hp		
Designation:	PBM-3/5		
Type:	Patrol flying-boat	Prop blades:	Curtiss

MARTIN PBM-3/5 MARINER — KNOWN SERIALS/IDENTITIES

Model	BuNo	Fuse marks	Nose marks	USCG/Navy Station
PBM-3	6582			Navy
PBM-3	6589	S-589	ASR-13	Navy
PBM-3	6613	S-613		Navy
PBM-3	6615	S615		Navy
PBM-3	6620			Navy San Francisco
PBM-3	6665	S-665		Navy
PBM-3	6666		FRISCO-D	Navy San Francisco
PBM-3	01663			Navy
PBM-5G	84686			USCG Salem, Massachusetts
PBM-5G	84728			USCG San Francisco
PBM-5G	84732		R-22	USCG Salem, Massachusetts
PBM-5G				USCG San Francisco

MARTIN P5M-1G/2G MARLIN

During May 1954 the US Coast Guard took delivery of the first of seven Martin P5M-1G Marlin twin-engined flying-boats powered by Wright Cyclone 18 compound R-3350 engines, with a top speed of almost 300mph and a maximum range of over 3000 miles. Two years later in 1956, four T-tail versions of the Marlin, designated P5M-2G, were delivered to the Coast Guard for use on long-range search and rescue (SAR) missions. All Marlins were returned to the US Navy during 1961 as they were plagued by high maintenance and operating costs. They were the last flying-boats operated in Coast Guard service.

The Martin Model 237 was the last operational flying-boat for the US Navy, being evolved from the World War II PBM-Mariner. It entered service during 1952 and was first line equipment until 1966. The design began in 1946 with a Mariner wing and upper hull combined with a new lower hull. The US Navy issued a contract on 26 June 1946 for a prototype and this, designated XP5M-1 flew on 30 May 1948. It was powered by

Parked at San Francisco USCG Air Station on its beaching gear is a Martin P5M-1G '1287' which was commissioned in September 1954. Seven of these early Marlins were operated by the service, it being the last flying-boat operated by the Coast Guard. *(Gordon S Williams)*

Wright R-3350 radial engines and was fitted with radar-operated nose and tail turrets, as well as a power-operated dorsal turret.

By July 1950, a US Navy contract authorised the P5M into production with the Glenn L Martin Company at Baltimore, Maryland. The first production Marlin flew on 22 June 1951. It differed from the prototype by being fitted with a large radome for the APS-80 search radar which replaced the nose turret. The dorsal turret was removed, and the flight deck raised to provide better visibility. Uprated engines in the form of R-3350-30WA were fitted in lengthened nacelles which were utilised as weapon bays.

Deliveries to the US Navy patrol squadrons commenced on 23 April 1952. Production of the P5M-1 Marlin totalled 114 by 1954. At least eighty subsequently became P5M-1S when fitted with AN/ASQ-8 magnetic anomaly detection (MAD) equiment, 'Julie' active echo sounding, 'Jezebel' passive sono-buoy detection and other equipment. Seven Marlins went to the US Coast Guard as P5M-1G, these later becoming P5M-1T trainers when the US Navy acquired them in 1961.

A major redesignation of the Marlin was produced in 1953, this having a T-tail, a lower bow chine line, improved crew accommodation and 3450 hp R-3350-32WA engines. The new P5M-2 was first flown during August 1953, going to the US Navy patrol squadrons on 23 June 1954. By the time production had ceased at the end of 1960, 145 P5M-2 Marlins had been built, including ten for the French Aéronavale. A P5M-2S version was

Four of the Martin P5M-2G Marlin 'T'-tail flying-boats were utilised by the USCG during 1961. Depicted is '1312' in flight. This later model of the Marlin served at St Petersburg Air Station, Florida, and San Francisco Air Station in California. *(USCG)*

equipped with the 'Julie-Jezebel' systems plus other new equipment.

Four Marlin P5M-2G flying-boats were delivered to the US Coast Guard in 1961, these serving at St Petersburg, Florida, air station and at San Francisco air station in California.

TECHNICAL DATA

Manufacturer:	Martin	Designation:	P5M-1G Marlin
Contract No.:	51-684, 52-982	Type:	Patrol flying-boat
Span:	118 ft	Unit cost:	Approx $1,500,000
Height:	31 ft	Length:	95 ft
Empty weight:	36,002 lb	Wing area:	1406 sq ft
Top speed:	250 mph	Gross weight:	70,000 lb +
Cruise:	150 mph	Range:	3000 miles plus
Engine:	Wright Cyclone 18 R-3350-30WA	Service ceiling:	24,000 ft
Take-off power:	2 × 2700 hp	Four-blade prop blades:	Hamilton Standard
Prop diameter:	15 ft 2 in		

TECHNICAL DATA

Manufacturer:	Martin	Prop diameter:	15 ft 2 in
Span:	118 ft 2 in	Designation:	P5M-2G Marlin
Height:	32 ft 9 in	Type:	Patrol flying-boat
Empty weight:	50,485 lb	Unit cost:	Approx $1,000,000
Crew:	11	Length:	100 ft 7 in
Top speed:	251 mph at sea level	Wing area:	1406 sq ft
Cruise:	150 mph at 1000 ft	Gross weight:	85,000 lb
Sea-level climb:	1200 ft per min	Range:	2050 st miles
Engine:	Wright Cyclone 18 R-3350-32WA	Service ceiling:	24,000 ft
Take-off power:	2 × 3450 hp	Four-blade prop blades:	Hamilton Standard

SERIAL INFORMATION

USCG BuNo	Commissioned	Decommissioned
P5M-1G 1284	May 1954	
1285	Jun 1954	
1286	Jul 1954	
1287	Sep 1954	Photo — San Francisco 30 Jul 1955
1295	Sep 1954	
1296	Jul 1954	Photo — San Francisco 27 Aug 1955
1297	Aug 1954	

MARTIN RM-1Z/VC-3A

The Martin 4-0-4 was a development of the earlier Martin Model 2-0-2 from which it differed primarily in having a 39-inch increase in fuselage length, was pressurised and had more powerful engines. The prototype Model 4-0-4 airliner was a modified Model 2-0-2 making its first flight on 21 October 1950. A total of one-hundred and one were built for use by the US airlines, deliveries commencing during the Autumn of 1951. Like the earlier Model 2-0-2, the Model 4-0-4 provided accommodation for a maximum of forty passengers.

With the order for two aircraft received from the US Coast Guard the total Martin 4-0-4 production reached one-hundred and three. This total included sixty for Eastern Airlines, and forty-one for Trans World Airlines. Deliveries were completed early in 1953.

Initially designated Martin RM-1Z two Martin 4-0-4 type VIP aircraft were delivered to the USCG in 1952. During 1962 the two transports were re-designated VC-3A under a DoD directive, being finally retired in 1969. Depicted on 18 November 1960 is RM-1Z '1283' in early USCG livery. *(USCG via William T Larkins)*

The two US Coast Guard Martin transports were initially designated RM-1 and given USCG serials '1282' and '1283' and were based with the Coast Guard air detachment located at Washington National Airport. They were used for administrative support of the USCG HQ and the Treasury Department until 1967, and then with the Department of Transportation, when it took over control of the Coast Guard.

It was during 1951 that the two Martin aircraft were ordered, initially receiving the designation RM-1, later RM-1G with the manufacturer's c/n 14290 and 14291. Having VIP executive interiors the designation was changed to RM-1Z. Under a Department of Defense Directive dated 6 July 1962, the designations allocated to the US Navy were standardised with those of the US Air Force variants and these two RM-1Z aircraft became VC-3A. The aircraft were retired during 1969, going to the US Navy with BuNos 158202 and 158203.

TECHNICAL DATA

Manufacturer:	Glenn L Martin	Prop diameter:	13 ft 1 in
Other designations:	RM-1Z	Designation:	VC-3A
Contract No.:	Tcg-38422	Type:	Personal transport
Span:	93 ft	Unit cost:	$647,140
Height:	28 ft 6 in	Length:	74 ft 7 in
Fuel:	1350 galls	Wing area:	846 sq ft
Empty weight:	27,800 lb	Oil:	50 galls
Crew:	2	Gross weight:	43,650 lb
Top speed:	312 mph at 14,500 ft	Passengers:	20
Cruise:	280 mph at 10,000 ft	Stall speed:	79 mph
Sea-level climb:	1790 ft per min	Range:	812 miles at 10,000 ft
Engine:	Pratt & Whitney R-2800-34 Double Wasp	Service ceiling:	27,600 ft
Take-off power:	2 × 2400 hp	Three-blade prop blades:	Hamilton Standard 23260

NAVAL AIRCRAFT FACTORY N3N-3

Four Naval Aircraft Factory N3N-3 trainers were acquired by the Coast Guard in a trade for four Grumman JF-2 Duck aircraft with the US Navy. These trainers provided a need in order to accelerate the Coast Guard pilot training programme. An extremely rugged aircraft known as the 'Yellow Peril' by the US Navy it was a curse and the salvation of thousands of fledgling aviators. Three of the N3N-3 trainers were commissioned during December 1940, with a rumour of a further three, but in January 1941 only one further trainer was received.

Brief mention must be made of the unique activities of the Naval Aircraft Factory (NAF) located in the Philadelphia Navy Yard, Pennsylvania. The N3N-3 was the last aircraft to be built by the NAF and the last biplane to serve with the US Navy.

Since 1916 the US Navy had actually been engaged in the design and manufacture of aircraft. A permanent manufacturing and test facility was considered desirable, this resulting in the NAF being authorised in 1917 and the factory was completed early the following year. In the immediate post-war years after World War I the NAF continued to manufacture outside designs, but this practice ended in 1922. Large-scale overhaul and modification work continued until World War II.

The factory became involved in significant manufacturing in the mid-1930s when it was decreed that the US Navy would build ten per cent of its aircraft. Among other things this allowed the US Navy to obtain accurate manufacturing cost data. By this time the US Navy had designed and built the prototypes of a new primary trainer, the XN3N-1. Rather than have it manufactured by the industry, it was put into production in Philadelphia, with deliveries commencing in 1936. From that date until the end of World War II, the NAF was again a major aircraft producer. A total of 817 N3N-1/3 trainers were built. Aircraft manufacture ended for good at the NAF at the end of World War II, and the former factory was then re-designed as the Naval Air Material Unit — NAMU.

When the US Navy developed a new primary trainer design in 1934, it was outwardly similar to the Consolidated NY-2/3 still in service and the Stearman NS-1 then under test in a civil prototype form. The major difference in the new N3N was its structure, which featured bolted steel tube fuselage construction with removable side panels for ease of inspection and maintenance. The wings were of all-metal construction, being fabric covered like the fuselage and tail. The prototype powered by a 220hp Wright J-5 radial engine was first flown in August 1935. Although the engine had been out of production since 1929, the US Navy had quantities in storage and wanted to use them up.

Testing of the XN3N-1 was successful at both the NAF at Philadelphia and Anacostia, both as a landplane and a single-float seaplane, so the US Navy put it into production. The first of

This Naval Aircraft Factory N3N-3 trainer 'V196' was the last of four delivered to the USCG, being commissioned in January 1941. It was an extremely rugged training aircraft and used in large numbers by the US Navy. At least one served at St Petersburg USCG Air Station during 1943/44. *(USCG)*

179 N3N-1s built was delivered in June 1936. Further development followed with a single prototype XN3N-2, while the fourth N3N-1 was modified as the XN3N-3. These two prototypes were powered with US Navy built versions of the 240hp Wright J-6-7, otherwise known as the R-769-96, engine.

Production of the N3N-1 was completed in 1938, to be followed by 816 N3N-3s, which differed outwardly from the N3N-1 in having a revised vertical tail shape and a single-strut landing gear. The earlier N3N-1s were delivered with a wide anti-drag ring around the engine, not fitted to the N3N-3 which served in primary training schools throughout World War II along with earlier models.

Early Coast Guard pilots were trained on the N3N-1 of the US Navy at NAS Pensacola, Florida. Unfortunately very little information is available on the history of the four Coast Guard N3N-3 trainers.

Aircraft history cards are always interesting, often revealing new information. The aircraft history card for N3N-1 BuNo 0056 reveals that between 8 March 1943 and 31 August 1944 this aircraft was assigned to Station Operations, Coast Guard air station St Petersburg, Florida.

TECHNICAL DATA

Manufacturer:	Naval Aircraft Factory	Take-off power:	1 × 235 hp
Span:	34 ft	Prop diameter:	9 ft
Height:	9 ft 3 in	Designation:	N3N-3
Fuel:	45 galls	Type:	Primary trainer
Empty weight:	2100 lb	Unit cost:	$20,868
Crew:	2	Length:	25 ft 7 in
Top speed:	126 mph	Wing area:	305 sq ft
Cruise:	72 mph	Oil:	3.8 galls
Sea-level climb:	860 ft per min	Gross weight:	2773 lb
Engine:	Wright Whirlwind 225 R-760.2	Range:	464 miles
		Service ceiling:	15,200 ft

SERIAL INFORMATION

USCG No	Commissioned	Decommissioned
V193	Dec 1940	
V194	Dec 1940	
V195	Dec 1940	
V196	Jan 1941	

NEW STANDARD NT-2

Two New Standard aircraft were among the fifteen miscellaneous aircraft handed over to the US Coast Guard during 1934. They had been confiscated for offences such as smuggling, violations of flying regulations or similar misconduct, being pressed into use by the US Customs Service. The Coast Guard only retained the two New Standard aircraft, as the rest were not considered usable and many were in a very poor shape.

During 1930 the US Navy had purchased six New Standard D-29s that had been modified to US Navy trainer specification and designated NT-1. Powered by a 100hp commercial Kinnier K-5 five-cylinder engine, the aircraft was of conventional design and initially did not have a US Navy designation. In its original form, the D-39/NT-1 featured a unique 'bath tub' cockpit in which the student and instructor sat in tandem in a single cockpit. The problems this arrangement caused, including drag and personal discomfort, soon resulted in conversion to the additional two cockpit configuration.

Rare photo depicting a New Standard NT-2 in Coast Guard livery, one of two seized from smugglers and put into use by the service during 1934. The engine was a 245 hp Wright J-6-7. *(Peter M Bowers)*

The two NT-2s acquired by the Coast Guard were in fact not derivatives of the NT-1, but much earlier built New Standard D-25A models, powered by the 245hp Wright J-6-7 engine, and used by the smugglers when they were captured. The first of these aircraft was allocated the USCG serial 311, later becoming V123 and was commissioned in March 1934, followed by the second aircraft which was commissioned in September 1934, being initially 312, later V124. Both crashed during the following year.

It is of historical interest to note that the New Standard design originated in Belgium as the Stampe-Vertongen D-29-A and was built under licence in the USA.

TECHNICAL DATA

Manufacturer:	New Standard	Prop diameter:	8 ft
Other designations:	D-25A	Designation:	NT-2
Span:	30 ft	Type:	Trainer
Height:	9 ft 7 in	Length:	24 ft 7 in
Fuel:	22 galls	Wing area:	245 sq ft
Empty weight:	1211 lb	Oil:	3 galls
Crew:	2	Gross weight:	1799 lb
Top speed:	99 mph	Stall speed:	47 mph
Cruise:	70 mph	Range:	280 miles
Sea-level climb:	480 ft per min	Service ceiling:	10,800 ft
Engine:	Wright J-6-7 Whirlwind	Two-blade prop blades:	Hamilton Standard
Take-off power:	1 × 225 hp		

SERIAL INFORMATION

USCG No		Commissioned	Decommissioned	
311	V123	Mar 1934	Oct 1935	Crashed
312	V124	Sep 1934	Nov 1935	Crashed

NORTH AMERICAN SNJ-5/6 TEXAN

During 1943 the first of a batch of fifteen North American SNJ-5/6 Texan single-engine training aircraft were delivered to the Coast Guard, the last of the consignment being received in 1945.

It was during 1937 that the US Army Air Corps organised a contest for a basic combat training aircraft, in a new category of trainer intended to have the same equipment and characteristics as operational aircraft. To conform to the requirement set out in the Army Air Corps Proposal 37-220, North American built a variant of their NA-16 Yale. Powered by a 600hp Pratt & Whitney R-1340 engine, it had a retractable undercarriage, provision for both forward and rear armament if required, navigation and engine instruments, all representative of a combat type.

This prototype was known as the NA-26 and became the immediate progenitor of the well-known and respected AT-6 Texan and Harvard types. It remained in production for a decade achieving wide fame and popularity. By the time production of the extensive series ended, North American had produced well over 16,000 examples of the same basic design, with many hundreds being built overseas.

The US Navy first ordered a version of the trainer late in 1936, with an initial contract for four NJ-1, known at the factory as NA-28. Power was by a 500hp Pratt & Whitney R-1340 Wasp engine, with the final batch in the order being temporarily fitted with a Ranger XV-770-4 inverted V-12 engine and designated NJ-2, later converted to NJ-1 standard. They served as training aircraft at NAS Pensacola, Florida, so could have been flown by USCG pilots under training.

An improved model known as the SNJ-1 (NA-52) was ordered by the US Navy in 1938, these having metal-covered rear fuselages. The batch only involved sixteen aircraft and they were assigned to the US Naval Reserve air bases. They never made it, being retained by the US Navy. Instead the Reserve received new SNJ-2 trainers from a batch of thirty-six ordered in 1939 with a further twenty-five more ordered in 1940, the two batches being designated NA-65 and NA-79 respectively. They had minor differences and the R-1340-65 engine distinguished the SNJ-2.

The US Navy placed their first order for the SNJ-3 in 1940, this featuring the triangular fin and rudder and blunt wingtips. Seventy were purchased on the US Navy contract. Minor improvements were incorporated in the 2400 SNJ-4s built in Dallas, Texas, and a further 1573 SNJ-5s differed only in having a 24-volt electrical system replacing the 12-volt system. These two versions were diverted from US Army Air Force contracts. A few SNJ-3J, SNJ-4C and SNJ-5C were fitted with arrester hooks and used for carrier deck-landing training.

Final version for the US Navy of the North American trainer was the SNJ-6 of 1944 which had strengthened wing panels and a redesigned rear fuselage. The US Army Air Force were the procuring agency for the 411 SNJ-6 trainers known as NA-121 by the manufacturer. The Texan operated by the US Navy remained in service in large numbers as primary, basic and instrument trainers. Those operated by the Coast Guard were retired by 1948, being returned to the US Navy with the loss of only one. In addition to being used for proficiency training, they also served on courier and administrative duties. One or two were assigned to each major USCG air station. Five Texans were used for cross-country navigation training. From the aircraft record cards the Texan was operated from Biloxi, Mississippi; Elizabeth City, North Carolina; San Francisco, California; Port Angeles, Washington; St Petersburg, Florida; and Salem, Massachusetts.

North American SNJ-5 Texan BuNo 90667, identical to the fifteen operated by the USCG during World War II. This well-known single engine basic trainer was operated by each major Air Station, and used on a variety of tasks, being finally retired during 1948. Only was one lost. *(USCG)*

TECHNICAL DATA

Manufacturer:	North American	Designation:	SNJ-5/6 Texan
Other designations:	NJ- AT-6	Type:	Basic trainer
Span:	42 ft	Length:	29 ft 6 in
Height:	11 ft 8 in	Wing area:	254 sq ft
Empty weight:	4158 lb	Gross weight:	5300 lb
Crew:	2	Range:	750 st miles
Top speed:	205 mph at 5000 ft	Service ceiling:	21,500 ft
Cruise:	170 mph at 5000 ft	Armament:	Two fixed .30in machine guns. One flexible .30in machine guns — rear cockpit.
Sea-level climb:	1200 ft per min		
Engine:	Pratt & Whitney R-1340-AN-1		
Take-off power:	1 × 550 hp		

SERIAL INFORMATION

USCG No		Commissioned	Decommissioned
SNJ-5	90669	Jan 1946	Salem Feb 1946 to May 1946
SNJ-5	90670	Jan 1946	
SNJ-5	90671	Mar 1945	Beloxi Apr 1945 to Jun 1946
SNJ-5	90672	Jan 1946	
SNJ-5	90673	May 1946	
SNJ-5	90674	Sep 1945	
SNJ-5C	90675	Jan 1946	
SNJ-5	90676	Jan 1946	
SNJ-5	90677	Apr 1946	San Francisco Jun 1946 to Jul 1946

NORTHROP RT-1 DELTA

Included in the nineteen aircraft procured by the US Coast Guard during 1935 was one which was possibly both unique and exotic. This was the Northrop Delta *Golden Goose* single-engine, low-wing landplane. It was a slick executive aircraft, which cruised at 185 knots having a service ceiling of 20,000 feet. Power was by a Wright Cyclone R1820-F3 engine fitted with a Hamilton Standard three-blade controllable pitch propeller. One of its tasks was that as command transport for the Secretary of the Treasury, and it was purchased under a Coast Guard contract for $41,909 although the figure of $45,000 appears in one document. It was an eight-place passenger transport and was used by Henry Morgenthau Jr, former Secretary of the Treasury at the time. It had a useful range of 1700 miles.

The Northrop Delta was developed in parallel with the Gamma series and was initially intended as a transport, differing from the first product of the Northrop Corporation in having a wider fuselage joined to Northrop Gamma wings. However, on the first day of October 1934, an amendment to the United States 1926 Air Commerce Act became effective. It required that for transport of passengers at night or over terrain not readily permitting emergency landings, the US airlines could use only multi-engined aircraft capable of flying with one engine inoperative.

The single Northrop RT-1 VIP staff transport '382' seen during take-off from Washington DC. It was delivered to the USCG in 1935, was later re-registered 'V150' and was decommissioned after an accident in 1940. It was both an exotic and unique type used by the Coast Guard. *(USCG)*

When Jack Northrop had undertaken the design of the Delta series, this stringent requirement was neither in force or anticipated. This effectively wiped out any potential market for this aircraft as no airline was prepared to purchase aircraft so restricted to daytime flying only and over certain areas. The military application was also very limited, so the Delta was soon restricted to the small executive market.

Northrop Delta s/n 74 was ordered by the Coast Guard and was designated RT-1, initially carrying the USCG serial 382 and, from October 1936, V150. The Delta ID-7 was delivered on 20 February 1935. After use by the Secretary of the Treasury it was operated as a USCG staff transport. Following an accident, the ID-7 was decommissioned in December 1940, being purchased in its damaged condition for $1400 by Charles Babb. The Delta, after being repaired, was licensed NC28663 and on 10 December 1941 was transferred to the ownership of Airmotive Inc. Its subsequent fate is unknown.

TECHNICAL DATA

Manufacturer:	Northrop	Take-off power:	1 × 735 hp
Other designations:	ID-7	Designation:	RT-1 Delta
Contact No:	Tcg. 23391	Type:	Executive transport
Span:	47 ft 9 in	Unit cost:	$45,000
Height:	10 ft 1 in	Length:	33 ft 1 in
Empty weight:	4540 lb	Wing area:	363 sq ft
Crew:	2	Gross weight:	7350 lb
Top speed:	219 mph at 6300 ft	Passengers:	8
Cruise:	200 mph	Range:	1650 miles
Sea-level climb:	1200 ft per min	Service ceiling:	20,000 ft
Engine:	Wright Cyclone SR 1820-F2	Three-blade prop blades:	Hamilton Standard controllable pitch

PIASECKI HRP-1 'FLYING BANANA'

During November 1948, the first of three Piasecki HRP-1 twin-rotor helicopters was purchased by the US Coast Guard at a unit cost of $256,912. Powered by a single Pratt & Whitney R-1840-AN1 engine, the helicopter known as 'Flying Banana' had a cruising speed of 64 knots and a range of 140 miles. Fitted with a rescue hoist, the HRP-1 carried a crew of two and could pick up eight survivors. They were all based at the air station located at Elizabeth City, North Carolina. Two were eventually returned to the US Navy, while one crashed in April 1951.

A Coast Guard Piasecki HRP-1 aptly named 'Flying Banana' seen parked at Elizabeth City Air Station, North Carolina, on 30 January 1951. It carries the US Navy BuNo 111821, was commissioned in November 1948, and returned to the US Navy in December 1951. Only three were used by the service. *(USCG via William T Larkins)*

The 'Flying Banana' design made use of the tandem-rotor layout favoured by designer Frank Piasecki. An early Piasecki Model PV-3 was ordered by the US Navy on 1 February 1944, but only one of the two XHRP-1 prototypes was accepted. An order for twenty HRP-1 helicopters followed. This model had a fabric-covered fuselage and three of these helicopters went to the US Coast Guard.

Piasecki Helicopter Corporation based at Morton, Pennsylvania, were successful in obtaining an evaluation and service trials for a batch of eighteen YH-21s for the US Air Force. These were based on the US Navy HPR-2 helicopter design — Piasecki Model PD-22, this being the first tandem helicopter delivered to the US Air Force. First flight was made on 11 April 1952. Trials were successful and thirty-two H-21A were ordered to serve with the Air Rescue Service, principally in the Arctic. With first flight during October 1953, the US Air Force purchased thirty-eight Piasecki Model 42 helicopters including six for Canada under the Mutual Defense Aid Programme.

Successive models included the H-21B for the US Troop Carrier Command as an assault helicopter, and the H-21C for the US Army and named Shawnee. There were also two H-21D helicopters powered by T58 shaft turbines. By this time the Piasecki company name had been changed to Vertol.

TECHNICAL DATA

Manufacturer:	Piasecki	Take-off power:	1 × 600 hp
Other designations:	Vertol 44. H-21	Designation:	HRP-1 Flying Banana
Rotor diameter:	41 ft	Type:	Rescue helicopter
Height:	14 ft 11 in	Unit cost:	$256,912
Fuel:	100 galls	Length:	83 ft 4 in
Empty weight:	5200 lb	Blade area:	135 sq ft
Crew:	2	Oil:	8 galls
Top speed:	98 mph	Gross weight:	6900 lb
Cruise:	74 mph	Passengers:	8
Sea-level climb:	650 ft per min	Hover ceiling:	5500 ft
Engine:	Pratt & Whitney R-1840-AN-1	Range:	140 miles
		Service ceiling:	8500 ft

SERIAL INFORMATION

USCG No	BuNo	Commissioned	Decommissioned	
1821	111821	Nov 1948	Dec 1951	Returned to US Navy
1823	111823	Nov 1948	Jan 1952	Returned to US Navy
1826	111826	Dec 1948	Apr 1951	Crashed

SCHRECK/VIKING OO-1

Listed only as 'seaplane experimental' in the Coast Guard register, USCG aircraft No.8 was imported from France and placed in Commission on 16 December 1931. This flying boat was very similar to models used by the French Aéronavale at the time, and was built by Hydravions Schreck-FBA Company of Argenteuil, France. The designer was Louis Schreck and many of his flying boats were in use in Canada. The one the Coast Guard purchased was designated Schreck FBA 17HT-4, a wooden-hulled pusher derived from the FBA designs of World War I.

Late in 1930 the Viking Flying Boat Company of New Haven, Connecticut, was formed to replace the previous Bourdon

Six of the Schreck/Viking OO-1 seaplanes were used by the USCG, the first being delivered in 1931, the remainder in 1936. Depicted is OO-1 'V152' which unfortunately crashed during March 1939. The beaching gear including the tail dolly is of interest. *(USCG)*

Aircraft Company, and they acquired the manufacturing rights to build the Schreck in the United States. As a result of Coast Guard experience with the single CG-8 — this being its initial USCG serial — five of these flying-boats were ordered from the Viking Company, and were built and delivered during 1936. These five flying boats were powered by a 250hp Wright Whirlwind R-760 radial engine and designated OO-1.

The single Schreck FBA-17HT-4 CG-8 was purchased during December 1931, receiving the early Coast Guard paint scheme. The last OO-1 flying-boat was decommissioned during April 1941. The OO-1 was an excellent craft for landing in an open sea, and they were operated from the Coast Guard air stations located at Biloxi, Mississippi, Cape May, New Jersey, Charleston, South Carolina, Miami and St Petersburg in Florida.

TECHNICAL DATA

Manufacturer:	Viking Boat Company	Prop diameter:	7 ft two-blade
Other designations:	V.2	Designation:	OO-1
Contract No:	Tcg. 26-271 14773	Type:	Flying-Boat
Span:	42 ft 4 in* 38 ft 7 in	Unit cost:	$7,215* $6,500
Height:	10 ft 4 in	Length:	29 ft 4 in
Fuel:	60 galls.	Wing area:	250 sq ft
Empty weight:	4200 lb	Gross weight:	5900 lb
Top speed:	104 mph	Stall speed:	48 mph
Cruise:	88 mph	Range:	390 miles
Engine:	Wright Whirlwind R-760	Service ceiling:	15,300 ft
Take-off power:	1 × 250 shp	Prop blades:	Wood-fixed pitch

Miscellaneous information:
*Refers to V107 Schreck FBA 17HT-4.

SERIAL INFORMATION

USCG No	Commissioned	Decommissioned	
V107	Dec 1931	Mar 1934	Destroyed by fire
V152	Oct 1936	Mar 1939	Crashed
V153	Oct 1936	Nov 1940	
V154	Nov 1936	Apr 1941	
V155	Dec 1936	Nov 1939	
V156	Dec 1936	Dec 1939	

SCHWEIZER RG-8A CONDOR

Two Schweizer RG-8A two-seat Condor aircraft have been transferred from the US Department of Defense (DoD) to the US Coast Guard and are based at Opa Locka airport, Florida on the inventory of the Miami USCG air station. They are employed in patrolling the Florida coastline on covert surveillance operations connected mainly with the drug interdiction conflict, and are a very unique aircraft type.

Constructed by the Schweizer Aircraft Corporation of Elmira, New York the SA 2-37A designated RG-8A by the DoD, is a modified version of Schweizer's SGM 2-37, designated TG-7A by the DoD, which is a motor glider developed for use as a special purpose aircraft for law enforcement, border surveillance and other selected military applications. The wings are of slightly greater span, with modified leading edges to improve stalling characteristics, and the SA 2-37A (RG-8A) has a much more powerful engine, fitted with mufflers — long exhausts on each side of the engine cowling — and driving a three-blade 'quiet' propeller. The aircraft is certified under Federal Administration Regulations Pt 23 for day and night VFR and IFRT operations, and has a very low acoustic signature. Requiring only 52hp from the 235hp 10-540 engine to maintain altitude in the quiet mission mode, it is said to be inaudible from the ground when flying in the 'quiet mode speed at 2000 — 3000ft or even lower.

Standard fuel capacity is more than trebled compared with the motor glider, and can be increased further by use of an optional auxilary tank. Its endurance in not listed in any known reference documents but is rumored as being eight hours. As one Chief of the USCG Aviation Division has remarked 'Pilots who drink coffee cannot fly the RG-8A.' Behind the new bulged cockpit canopy is a 65-cubic foot payload bay designed to accept various sensor pallets such as low-light level TV, infra-red imaging systems, standard cameras, or other payloads specified by customers. The pallet can be removed quickly for replacement or maintenance, and a removable underfuselage skin and rear hatch doors provide access to the entire rear section of the aircraft. Other power plants and larger payload versions of the SA 2-37A are possible both for specific surveillance applications and for such alternative roles as basic and advanced training, mission operator training glider, banner towing and priority cargo delivery.

HISTORY

First flight of the SA 2-37A prototype N9237A, a modified SGM 2-37, was made in 1986. Three RG-8As 85-0047, 85-0048 and 86-0404 were acquired by the US Army. The second aircraft 85-0048 was lost in an accident, the other two being now

As part of its ever increasing role in drug interdiction the US Coast Guard operates two Schweitzer RG-8A reconnaissance surveillance aircraft. Depicted is 86-0404 now registered '8102' in the Coast Guard inventory and is seen in flight over the Caribbean. Despite its size the RG-8A carries a vast array of avionics and communication equipment. *(USCG)*

transferred to the US Coast Guard. The prototype was fitted with a Hughes AN/AAQ-16 thermal imaging system.

The Schweizer SGM 2-37 two-seat motor glider was designed for the US Air Force under the designation TG-7A and eleven were delivered to the US Air Force Academy located at Colorado Springs, Colorado. This motor glider has an endurance of two hours thirty minutes.

AVIONICS & EQUIPMENT

Avionics included in the Schweizer RG-8A Condor two-seat special missions aircraft are a Litton inertial navigation system, standard Bendix/King avionics comprising dual KX155 nav/com — one with VOR/ILS, one with VOR indicator, and single KG258 artificial horizon, KEA129 encoding altimeter, KR87 ADF with indicator, KT76 transponder, KR21 marker beacon receiver, KMA24H audio control panel, and dual engine mufflers with resonator quiet kit. Optional avionics and equipment include KRA 10A radar altimeter, LORAN system, and KFC 150 autopilot requiring KCS 55A slave compass.

The RG-8A Condor can be fitted with armour seat protection for the two pilots who sit side by side, and lighting compatible with night vision goggles. The two USCG RG-8A aircraft are powered by a 235hp Textron Lycoming 10-540W3A5D engine. Both initially retained their original US Department of Defense

(DoD) serial numbers and are finished in low visibility markings, carry a small US national insignia, 'COAST GUARD' on either side of the fuselarge and 'USCG' on the starboard upper wing.

When the US Army acquired two examples of the two-seat Schweizer SA 2-37A, developed from the SGM 2-37 motor-glider, and designated RG-8A in 1986, they were heavily involved in the continued investigation into quiet surveillance aircraft that had begun with the Lockheed QT-1 — itself a modified Schweizer SGS 2-32 sailplane — and continued with the Lockheed QT-2PC and YO-3A aircraft.

After the loss of the RG-8A 85-0048, another RG-8A — 86-0404 — replaced it.

TECHNICAL DATA

Manufacturer:	Schweizer Aircraft Corporation	Take-off power:	1 × 235 hp
Other designations:	SA 2-37A	Prop diameter:	7ft 2 in
Span:	61ft 6 in	Designation:	RG-8
Height:	7ft 9 in	Type:	Law enforcement surveillance
Fuel:	52 US gals (67 US gals)	Length:	27ft 9 in
Empty weight:	2025 lb	Wing area:	199.4 sq ft
Crew:	2	Gross weight:	3500 lb
Cruise:	75% power 159 mph 65% power 148 mph at 5,000ft	Stall speed:	80/92 mph
		Service ceiling:	18,000 ft
Sea-level climb:	960ft per min.	Prop blades:	McCawley three-blade constant speed
Engine:	Textron Lycoming 10-540-W3A5D		

SERIAL INFORMATION

USCG 8101	ex-USAF 85-0047
8102	86-0404

SIKORSKY S-39

During 1929 there were plans to replace the five existing aircraft in use with the US Coast Guard with a lesser number of better suited aircraft. It was 1931 before the US Congress made an additional appropriation which enabled the Coast Guard to acquire a number of flying-boats best suited to its needs.

However, during 1930, a single Sikorsky S-39 twin-boom amphibian was loaned by Sikorsky to the Coast Guard for aerial patrol of the International Yacht races, which commenced on 13 September 1930, off Newport, Rhode Island. Records show that the S-39 was retained by the USCG until 1935.

During 1934 the US Customs Services turned over fifteen miscellaneous aircraft to the Coast Guard. These had been confiscated for offences such as smuggling, violation of flying regulations or similar misconduct, and were initially pressed into use by the US Customs Service for border patrols. The list included a single Sikorsky S-39. Most of these fifteen aircraft were soon destroyed as they were obsolescent.

Igor I Sikorsky designed an interesting family of early amphibians in the 1920s for the sportsman, executive owner pilot, these including the S-34, S-36, S-38, S-39 and S-40. They were all basically of the same configuration, with the wing high above the fuselage in parasol fashion by struts and the empannage held in position by outrigger supports.

The Sikorsky S-39 first flew on 24 December 1929, but arrived at a very bad time. The economics of the world in general were such that no sane person would be likely to run out and spend $20,000 on a new aircraft. The S-39's first flight was just two months to the day after the huge Wall Street crash on 24 October 1929, which climaxed four days later.

This new amphibian was Igor Sikorsky's first all-metal aircraft, and also the first Sikorsky aircraft to be built at Connecticut. The prototype is reputed to have been the first aircraft for which there was a full scale mock-up built prior to construction of the actual aircraft. Despite the S-39 appearing at such a critical time, preventing mass production (only twenty-three aircraft including prototypes were built) it nevertheless established an enviable record.

Exploration flights involving 60,000 miles of flying in Africa, were followed by a similar expedition in Borneo, while another S-39 explored Australia. One S-39 held the International Altitude Record for Class C3 aircraft of 18,641.676 feet established on 29 January 1935. On 10 February 1935, the same S-39 set an International Speed Record for Class C3 aircraft over a 1000 kilometre (621 mile) course, with an average speed of 99.950 miles per hour. This was held by the S-39 until 9 May 1937. However, it was not all glory for the Sikorsky S-39. The prototype crashed on its third flight, several were scrapped, one was made into a boat, two were smashed by a drunken fisherman and one by a drunken driver.

The US Army Air Corps operated the only S-39 to be procured by a US military service on a permanent basis. It was c/n 919 with serial 32-411, designated YIC-28 and delivered to Wright Field about August 1932, being transferred to West Point in November of that year where it was possibly used for familiarisation flights. It was apparently damaged beyond repair in a landing accident on the Hudson River early in October 1934. This aircraft was last listed by the US Army Air Corps in a report issued on the last day of October 1934.

During 1939, Columbia Pictures made a film called *Coast*

Rare photo depicting a Sikorsky S-39 twin-boom amphibian in US Coast Guard early markings. It was loaned to the service by Sikorsky during 1930 and retained until 1935. During 1939 an S-39 was used in the film *Coast Guard. (USCG via William T Larkins)*

Guard starring the well-known film actors Randolph Scott, Ralph Bellamy and the charming Frances Dee. Unfortunately no aircraft from the existing Coast Guard fleet were used, the types involved carrying fictitious serial numbers, but full USCG livery and markings. These film aircraft were based at Burbank Airport, California, the air terminal for Hollywood and its stars.

Known aircraft types used in the film included an American Eagle single-engined biplane carrying the serial 504, a Stearman C3 biplane with the serial 600 and a single Sikorsky S-39 flying-boat amphibian ex-NC-52V c/n 914. It apparently carried no serial number.

Coincidental that when the US became involved in World War II this S-39 NC-52V was pressed into rescue service with the Civil Air Patrol. It served with Civil Air Patrol Base 10 at Beaumont, Texas. On Armistice Day 1942 a CAP operated Fairchild 24 NC16814 had ditched into the sea at 1640 GMT during a routine patrol. The waves were running 20/30ft high. The S-39 was launched and guided by a rescue vessel arrived at the wreckage of the Fairchild at Lat 29'20° Long 93'22° and landed at 1752 GMT. Landing was next to a suicide but pilot Robert F Neel made it, smashing a wing float in the process. The downed Fairchild crew were recovered but Pilot Neel was unable to restart the S-39 engine. A US Coast Guard cutter arrived shortly after and attempted to tow the Sikorsky to port but it sank at 2250 GMT.

Despite having the most detailed history of all the Sikorsky S-39 amphibians built, it is difficult to pinpoint the S-39 loaned to the US Coast Guard in September 1930, and rumored to have remained in USCG service until 1935. We assume, repeat assume, that the amphibian was none other that X963 c/n 1, actually the second S-39 to be built, which was retained by the Sikorsky company as a utility aircraft. There is no doubt that the type was extremely strong, its handling qualities were described as excellent, and water characteristics were outstanding. It landed slowly and beautifully and take-offs were clean. Stable in flight in all directions and very very pleasant to fly, the rugged dependability of this little amphibian will always be remembered with fondness.

TECHNICAL DATA

Manufacturer:	Sikorsky	Designation:	S-39
Other designations:	C-28	Type:	Amphibian flying-boat
Span:	52 ft	Unit cost:	$21,500 later $17,500
Height:	12 ft	Length:	32 ft 6 in
Empty weight:	2678 lb	Wing area:	350 sq ft
Crew:	2	Oil:	6 galls
Top speed:	119 mph	Gross weight:	4000 lb
Cruise:	100 mph	Passengers:	2
Sea-level climb:	750ft per min.	Stall speed:	54 mph
Engine:	Pratt & Whitney	Range:	400 miles
	Wasp Jr R-985	Service ceiling:	18,000 ft
Take-off power:	1 × 300hp	Prop blades:	Hamilton Standard

SIKORSKY HNS-1 HOVERFLY

The helicopter has advanced a long way since its beginning, so much so that even those who have been participants in rotary-wing progress are more than impressed by its achievements. From the awkward, early means of vertical flight to the fast, efficient and powerful helicopters today, this astounding vehicle has only been limited by the imagination of the dedicated people who make up the helicopter industry.

Tens of thousands of rescues that have already been performed by the helicopters of many operators, including the US Coast Guard, represent one of the brightest and most romantic chapters in the history of flight. Aside from the tremendous human value, the many rescues demonstrate the unique ability of the helicopter to render valuable service under circumstances and conditions where no other vehicle could be used.

In the hands of US Coast Guard aviators, helicopters have saved thousands of lives in the years since November 1943, when the first Sikorsky HNS-1 was delivered to the USCG air station at Floyd Bennett Field, Brooklyn, New York. From the very beginning of helicopter development, the service has shown its awareness of the potential value of rotary-wing aircraft in the performance of Coast Guard functions and by 1941 the USCG was seriously interested in developing the the helicopter for search and rescue (SAR).

On 14 September 1939, Igor I Sikorsky lifted the wheels of his famous VS-300 helicopter just inches off the ground at Stratford, Connecticut. This brief flight, witnessed by a few engineers and mechanics, proved to be an aviation milestone, leading to a series of helicopter advances that made the helicopter for the first time an instrument of practical use to mankind. Four years later in 1943, the helicopter industry and the world's first helicopter production line at Bridgeport, Connecticut, were established.

EARLY INTEREST

The US Coast Guard, then under World War II US Navy jurisdiction, was one of the first of the US services to recognise the potential of the helicopter. During April 1942, after witnessing early flight tests of the Sikorsky XR-4, a test model built for the US Army Air Force, Commander W A Burton, Commanding Officer of the US Coast Guard air station at Brooklyn, New York, wrote in his report: 'The helicopter in its present stage of development has many of the advantages of the blimp and few of the disadvantages. It hovers and manouvres with more facility in rough air than the blimp. It can land and take-off in less space. It does not require a large ground-handling crew. It does not need a large hangar. There is sufficient range — about two hours — in this particular model to make its use entirely practical for harbor patrol and other Coast Guard duties.'

On the basis of this recommendation, Commander F A Leamy, Chief of Aviation Operations, Coast Guard headquarters, in June 1942, advocated purchasing several Sikorsky helicopters for training and experimental development.

The Vought Sikorsky Model 316A was the first helicopter produced for the US armed forces in other than experimental quantities. Designated the R-4, a contract for the development experimental helicopter — XR-4 was awarded to the Vought Sikorsky Division of United Aircraft in 1941. The XR-4 design retained the single-rotor layout of the VS-300 but introduced a faired fuselage with side-by-side seating for the crew of two and full dual controls.

It was powered by a 165hp Warner R-500-3 engine mounted in the fuselage of the XR-4, which drove both the main rotor and the anti-torque tail rotor through transmission shafts and gear boxes. The flying controls set a standard from which there has been little deviation in subsequent helicopters, and comprised the cyclic pitch leaver, which controlled variation of the pitch of the main rotor blades in the course of a single revolution for forward or backward flight, the collective pitch lever, varying the pitch of all main rotor blades simultaneously for vertical flight control, and the rudder pedals controlling the pitch of the tail rotorblades.

Making its first flight on 14 January 1942, the XR-4 41-18874 was subsequently flown in early stages from Stratford, Connecticut, to Wright Field, Dayton, Ohio, thus accomplishing the world's first long-distance helicopter cross-country flight. During 1943, it was re-engined with a 180hp Warner R-550-1 engine and the original 36-ft diameter main rotor was replaced by one of 38-ft diameter. The designation was changed to XR-4C.

For further service trials, the US Army Air Force ordered thirty R-4s of which the first three were designated YR-4A and the remainder YR-4B. These models included the 38-ft diameter rotor and the 180hp R-550-1 engine. Gross weight was increased from the 2,450 lb of the prototype to 2,900 lb, but the top speed fell from 102 mph to 75 mph. The Sikorsky YR-4B helicopter were extensively tested under varied World War II conditions, including tropical trials in Burma and winterization trials in the cold of Alaska. During May 1943 after trials with an YR-4B operating from a small platform on a tanker, three Sikorsky helicopters were allocated to the US Coast Guard.

USCG HNS-1

Following a US Navy decision to procure helicopters for evaluation, made in July 1942, action was taken later in 1942 to obtain a single Sikorsky YR-4B from the USAAF, with the designation HNS-1 and the US Navy BuNo 46445. The US Navy accepted this first helicopter on 16 October 1943, with the second — BuNo 46699 — and third — BuNo 46700 — following

Helicopter pioneer Cdr F A Erickson, USCG, flies this Sikorsky HNS-1 BuNo 39040 whilst a crewman demonstrates the use of the U-type harness for air/sea rescues. The demonstration took place in Jamaica Bay on 25 August 1944, the USCG personnel being from the Air Station at Brooklyn, New York. *(USCG)*

on 2 November, 1943. Events moved swiftly during November, and by the 19th the Coast Guard Air Station at Floyd Bennett Field, Brooklyn, New York, was officially designated as a helicopter training base with the three Sikorsky HNS-1 helicopters. In order to train helicopter pilots twenty-two more HNS-1s were acquired later with BuNos 39033 to 39052, 75727 to 75728, the first twenty of these being ex-USAAF R-4Bs. These were powered by a 200hp Warner R-550-3. engine.

Coast Guard records indicate that in the second half of 1944 twenty-one HNS-1 helicopters were acquired, these being returned to the US Navy in late 1947 and early 1948, except for two which crashed during December 1946.

Meanwhile at the Coast Guard air station at Brooklyn, a special movable platform 40 feet by 60 feet had been developed which could simulate the motion of a ship's deck at sea. Further development work on stretchers, slings and related equipment progressed rapidly. One of the many people involved in this work was Igor Sikorsky's son, Sergei, who was serving in the Coast Guard at the time. He was crew chief on one of the original HNS-1s and spent nearly two years working with other military personnel on the development and perfection of the helicopter rescue hoist. On the first day of May, 1944, in HNS-1 BuNo 39040 the first helicopter hoist of a man took place at Brooklyn air station, New York. The pilot of the helicopter was Commander Frank E Erickson, and the crew man hoisted was Sergei Sikorsky. This was the first Coast Guard helicopter to be fitted with a hoist, and is now on display in the Sikorsky museum located st Stratford, Connecticut.

Commander Frank A Erickson USCG worked closely with Igor Sikorsky in pioneering development of the helicopter as a rescue vehicle. Frank Erickson was the first US Coast Guard pilot to qualify as a helicopter pilot and it is recognised he pioneered rotary wing development for military use. During World War II all Allied helicopter pilots were trained by the Coast Guard at Brooklyn. Coast Guard personnel trained British pilots who undertook a joint British-American helicopter trial on board the merchant ship *Daghestan* which was fitted with a landing deck and carried two Sikorsky HNS-1 helicopters. The open sea trials on board the *Daghestan* took place on 29 November 1943, off Bridgeport, Connecticut, was 328 landings made by British and American pilots.

Enthusiasm ran high following these first flight tests. Action was taken to provide a second ship, the *Governor Cobb*, a merchant vessel, to be operated by the Coast Guard. This old coastal passenger ship had been converted into the world's first helicopter carrier. On 29 June 1944, Commander Frank Erickson made the first landing on its deck in Long Island Sound. As the war progressed and the U-boat threat moved deeper into the North Atlantic and then abated, the USCG re-orientated its helicopter research from anti-submarine warfare to that of search and rescue (SAR). It was Frank Erickson who pioneered this Coast Guard activity, developing much of the rescue equipment himself and carrying out the first life-saving flight himself. He delivered two cases of blood plasma lashed to the floats of a HNS-1 following the explosion on board the destroyer USS *Turner* off Sandy Hook, New Jersey, on 3 January 1944, in the Ambrose Channel. He airlifted the blood plasma from Manhattan's South Ferry. As a result of this flight, further studies explored the helicopter's possibilities as a flying ambulance.

Following the tropical hurricane of 14 September 1944, four HNS-1 helicopters assisted in making a storm survey at Moriches, Long Island, and were found well suited for disaster survey work. It was at this time that the rescue harness was combined with the US Navy regulation life jacket. A demonstration of the new rescue harness was held off Manasquan, New Jersey, on 3 October 1944.

The first actual use of the hydraulic winch hoist in a rescue occurred on 29 November 1945, during a severe storm in the vicinity of Bridgeport, Connecticut. Two crew members of a barge, which had broken away from its tow and was in imminent danger of breaking up, were successfully rescued by a hoist-equipped HNS-1 helicopter sent out from the Sikorsky factory.

During December 1945, tests were conducted at Brooklyn air station carrying a stretcher under the fuselage of a Sikorsky HNS-1 at a level about four feet below the bottom of the floats.

Seen during 'Operation Highjump' during 1946/47 in the Antarctic and operating from the USCG icebreaker *Northwind* is a Sikorsky HNS-1 BuNo 39043. This early helicopter was pioneered by Commander Frank Erickson, veteran USCG pilot. Some twenty-one were believed to have been used by the service between 1943/48. *(AP Photo Library)*

No difficulties were encountered under all condition of hovering and cruising flight, nor was it found hazardous to land astride the stretcher. Further development work on stretchers, slings and other related equipment progressed rapidly.

On 1 April 1944, the first civilian rescue by helicopter occured. In a routine patrol training flight from Brooklyn, Lieutenant (jg) W C Bolton USCG sighted and rescued a boy from a sand bar after he had been marooned there by the tide off Jamaica Bay, New York. Bolton served at the Brooklyn air station from 1942 to 1947. He won the Air Medal for his role as a pilot of a Sikorsky HNS-1 helicopter in the rescue operation involving a Belgian Sabena Douglas DC-4 airliner that had crashed at Gander, Newfoundland. The records at the Brooklyn air station show that in the first quarter of 1945 there were 147 assistance flights by USCG helicopters, during which a total of 1,123 miles were flown.

Spraying tests were conducted with a Coast Guard HNS-1 helicopter BuNo 39052 in cooperation with the US Bureau of Entymology and Plant Quarantine, US Department of Agriculture on 17 May 1946, to determine the effectiveness of the helicopter as a means of applying insecticides to wide areas. Today the spraying capability of the helicopter is a valuable tool in agriculture throughout the world. During 1946/47 the US Navy Antarctic project Operation High Jump with thirteen ships of all sizes, 4000 personnel and twenty-six aircraft, the largest Antarctic expedition ever organised, took place. The air armada included one US Coast Guard HNS-1 helicopter BuNo 39043 which operated from platforms on the forward deck of the USCG icebreaker *Northwind*.The helicopter, along with others from the US Navy, was used largely for ice reconnaissance while in the ice pack and for seeking out ice-free areas for the Martin PBM Mariner flying boats to take-off and land.

During September 1946, when the Belgian airliner crashed in Gander, Newfoundland, the crash site was inaccessible to all but the newly-developed helicopter rescue vehicle. With the US Coast Guard air station at Brooklyn over 1000 miles away, a Sikorsky HNS-1 was disassembled, flown to Gander from Brooklyn in a Douglas C-54 Skymaster transport of the US Army Air Force, re-assembled and flown to rescue eighteen survivors. They were airlifted to a Coast Guard PBY Catalina and then flown to hospital. The total time from notification at the Brooklyn Operation Room, to the survivors' arrival at hospital was less than forty-eight hours. This item represents only a spattering of the colorful history of both the Brooklyn air station and the early Sikorsky HNS-1 helicopter.

Frank Erickson developed the idea and the techniques of power hoist equipment for practical use in helicopters. He demonstrated this in Jamaica Bay in 1944 as pilot of the first helicopter (HNS-1) pickup of a man on 11 August the first pickup of a man floating in water on 14 August and the first pickup of a man from a life raft on 25 September. These demonstrations led to an official commendation which he received in February 1945. His techniques in the use of the hydraulic hoist and related lifesaving equipment proved invaluable to military services and to non-military in the use of the helicopter for rescues, lifting of personnel, equipment, and cargo among other uses. His early demonstrations convinced the US Army Air Forces to use that equipment overseas and influenced the design of numerous machines in their development stages.

Captain F A Erickson USCG later developed a flight stabiliser for helicopters and inflatable pontoons for landing helicopters on water. In June 1946, Erickson was assigned to the US Coast Guard air station located at Elizabeth City, North Carolina, where the training and equipment, including the HNS-1, were moved from Brooklyn because of better flying conditions. Consequently the USCG Rotary Wing Development Project Unit was established at Elizabeth City.

It was a tentative beginning. But the awkward, flimsy, vertical-lift vehicle of the 1940s showed the promise of great accomplishments to come. And the faith and persistence of these US Coast Guard Aviation personnel who evaluated its initial trials were to open new horizons in man's conquest of the air.

TECHNICAL DATA

Manufacturer:	Sikorsky
Other designations:	R-45, VS-316A2
Rotor diameter:	38 ft
Height:	12 ft 5 in
Fuel:	30 galls
Empty weight:	2020 lb
Crew:	2
Top speed:	75 mph
Cruise:	60 mph
Sea level climb:	780 ft per min
Engine:	Warner R-550-3
Take-off power:	1 × 200 hp
Tail-rotor diameter:	8 ft 2 in/three
Designation:	HNS-1
Type:	General purpose helicopter
Unit cost:	$43,940
Length:	48 ft 4 in
Blade area:	651/2 sq ft
Oil:	3 galls
Gross weight:	2600 lb
Range:	65 miles
Service ceiling:	8200 ft

Miscellaneous information:
US Army Air Force Spec: X417-8
US Navy Spec: SD-364-1

SERIAL INFORMATION

BuNo.39033 ex-USAAF	YR-4B	43-28227
39036	R-4B	43-46507
39037	R-4B	43-46512
39040	R-4B	43-46525
39042	R-4B	43-46543
39043	R-4B	43-46544
39045	R-4B	43-46552
39051	R-4B	43-46586
39052	R-4B	43-46595
46445	YR-4B	42-107239

SIKORSKY HOS-1

The Sikorsky R-6 helicopter was developed in paralled with the R-5, being the Sikorsky Model 316B, and was a refinement of the earlier pioneer R-4 known as the Hoverfly, having the same rotor and transmission system with an improved fuselage. One XR-6 prototype was built by Sikorsky powered by a 225hp Lycoming 0-435-7 engine. Nash Kelvinator built twenty-six similar YR-6As followed by a production run of 193 R-6As. These were all built for the US Army Air Force.

The US Navy procured three Sikorsky XR-6As from the US Army Air Force in the latter part of 1943, the designation XHOS-1 being assigned. First acceptance was during September 1944, the helicopter going to the US Coast Guard at Floyd Bennett Field for evaluation, joining the R-4 or HNS-1. A batch of thirty-six HOS-1s followed having the BuNos 75589-75624, these also being acquired from the US Army Air Force R-6A production. All were produced by Nash Kelvinator under licence from Sikorsky.

Twenty-seven Sikorsky HOS-1 helicopters were purchased by the USCG after evaluation at Floyd Bennett Field, New York. Depicted is USCG HOS-1 BuNo 23470 seen making a landing on an improvised landing pad in Newfoundland in September 1946. It rescued survivors from a crashed Douglas DC-4 airliner. *(USCG)*

Between January 1945 and January 1946, the US Coast Guard purchased twenty-seven of the R-6A type, designated HOS-1. They all retained the US Navy BuNos in two groups. Powered by the Franklin 0-405-9 engine, they had a cruise speed of 75 knots, a range of 245 miles and could carry three people.

These helicopters were returned to the US Navy by the Coast Guard between May 1947 and May 1949, except for two which crashed in November 1946 and May 1947 respectively. One HOS-1 fitted with floats joined a HNS-1 in the shipboard trials on the merchant ship *Daghestan* and both types were involved in helicopter pilot training at Brooklyn USCG air station. From a photo one that crashed at Brooklyn was BuNo 75595.

TECHNICAL DATA

Manufacturer:	Sikorsky	Take-off power:	1 × 235 hp
Other designations:	R-6	Tail rotor diameter:	7 ft 10 in
Rotor Diameter:	38 ft	Designation:	HOS-1
Height:	10 ft 5 in	Type:	Utility helicopter
Fuel:	67 galls	Unit cost:	$47,800
Empty weight:	2252 lb	Length:	48 ft
Crew:	2	Blade Area:	65 sq ft
Top speed:	96 mph	Oil:	4 galls
Cruise:	90 mph	Gross weight:	3243 lb
Sea-level climb:	620 ft per min	Range:	245 miles
Engine:	Franklin 0-405-9	Service ceiling:	10,200 ft

SERIAL INFORMATION

BuNo. 75595	ex-USAAF	R-6A 43-45415
75610		R-6A 43-45531

SIKORSKY HO2S-1/HO3S-1G

So successful were the development and early flight trials of the Sikorsky R-4 (HNS-1) training and Coast Guard operational helicopter that the US Army Air Force issued a specification for a rather larger type to fit a requirement for an observation machine. This resulted in the Sikorsky Model 327, which was designed to fit the new specification following closely the general layout of the R-4, with main and tail rotors shaft-driven from the engine in the fuselage. However, in detail it was a completely new design.

Early in 1943 four prototypes were ordered with the designation XR-5, with a fifth to follow. First flight was on 18 August 1943, at Bridgeport, Connecticut. Powered by a 450hp Pratt & Whitney R-985-AN-5 engine, the XR-5 rotor diameter was forty-eight foot and gross weight 4850lb. The crew of two sat in tandem. Two of the five prototypes were later fitted with British equipment and designated XR-5A.

In preparation for intensive service trials, the US Army Air Force ordered twenty-six YR-5As, similar to the prototypes, and 100 R-5A models of which only the first thirty-four were built. These had provision for stretcher carriers on each side of the fuselage, and were the very first helicopters to be used by the Air Rescue Service. Later twenty-one R-5A helicopters became designated R-5D when they were modified to carry a second passenger and fitted with a rescue hoist, an auxiliary external fuel tank, and a nose wheel. Five YR-5A were modified to YR-5E standard having dual controls.

The two R-5A helicopters, redesignated HO2S-1 for service with the Coast Guard with US Navy BuNos 75689/90 were

A Sikorsky HO3S-1 '232' of the USCG demonstrates the use of wheel floats for use in an emergency landing on water at Elizabeth City Air Station, North Carolina, during April 1951. The type was delivered to the service during 1946, a total of nine being procured. *(USCG)*

actually ex-YR-5A machines. The first was ex-USAAF YR-5A/YR-5E 43-46613 which went to the USCG in January 1946, crashing during October 1950. The second was ex-YR-5A 43-46618 which went to the USCG in January 1946, being disposed of a few months later in June. Cruise speed was 80 knots and it had a range of 230 miles.

The evaluation of these early Sikorsky helicopters by the US Coast Guard emphasises not only the interest the service had in the value of the helicopter, but it illustrated the skill and experience the Coast Guard aviators had built up with the earlier pioneer helicopter, the HNS-1. Trials were held at Floyd Bennett Field, the home of Brooklyn USCG air station during December 1945. It was possibly the first tri-service — US Army Air Force, US Navy, US Coast Guard — evaluation exercise involving an operational military helicopter. These early Sikorsky R-5 helicopters had a tail wheel.

During June 1948, the US Air Force helicopters were re-designated with the 'H' instead of 'R' prefix designation. The helicopter was in service with the Air Rescue Service and to expand this valuable work, thirty-nine were purchased in 1948 designated H-5G and fitted with rescue hoist and other equipment. Sixteen Sikorsky H-5Hs were purchased during 1949 having a combination of a wheel/pontoon undercarriage and a gross weight of 6500lb.

The US Navy placed production orders for a four-seat version of the Sikorsky helicopter designated HOS3-1 and this was equivalent to the US Air Force H-5F. It served in a wide variety of roles giving outstanding service with HU-1 squadron in the Korean conflict. At least eighty-eight were procured.

During 1946 the Coast Guard procured nine Sikorsky HO3S-1G helicopters which were powered by Pratt & Whitney Wasp Jr R985-AN-5 engine. They were fitted with a rescue hoist. Some USCG models were equipped with inflatable flotation bags. The first HO3S-1G delivered to the Coast Guard was '233' during August 1946, with the last and ninth '238' taken on charge in February 1950. The type stayed in Coast Guard service until 1957, with four of the nine involved in crashes.

OPERATIONS

On 20 January 1950, the schooner *Gee Gee III* went aground off False Cape, Virginia, with five men on board. The difficult task of refloating the yacht finally took a week to accomplish, involving the united efforts of the USCG cutters *Marion* and *Cherokee*, the Virginia Beach and Caffeys Inlet Coast Guard stations, and constant air patrol during the operation by a Sikorsky HO3S-1G helicopter from the Elizabeth City Air Station, North Carolina.

During the two-day American Helicopter Society Show held at Anacostia Naval Air Station, and Haines Point, Washington DC, on 28/29 April 1951, US Coast Guard Sikorsky HO3S-1G helicopters gave a demonstration. This included the use of inflatable floats for emergency water landing, hoist pick-ups, etc. The helicopters were from the Elizabeth City air station.

At 0907 on 24 July 1952, the Coast Guard were notified that appendicitis victim Maurice A Piggins on board the fishing vessel *Antonio* 160 miles east of Chatham, Massachusetts, needed to be removed to hospital. A Martin PBM-5G Mariner of the Coast Guard piloted by Lieutenant John Vukic landed alongside the *Antonio* at 1217 and returned to the USCG air station at Salem, Massachusetts at 1340. A Sikorsky HO3S-1G helicopter 235 piloted by Lieutenant Commander George W Girdler, airlifted the patient to the US Public Health Hospital located at Brighton, Massachusetts.

During the early winter months of 1954 a Coast Guard Sikorsky HO3S-1G helicopter 233 was based at the Moorings, East 9th Street, Cleveland, Ohio, for approximately two weeks for ice observation and search and rescue (SAR) while the Coast Guard cutter *Kaw* was on temporary duty (TDY) breaking ice in Buffalo Harbor. The USCG helicopter was piloted by Lieutenant Robert W Bruck, assisted by Aviation Mechanic Third Class George E Fennell.

TECHNICAL DATA

Manufacturer:	Sikorsky	Take-off power:	1 × 450 hp
Other designations:	S-51, H-5	Tail-rotor diameter:	8 ft 5 in three blade
Rotor diameter:	49	Designation:	HO2S-1, HO3S-1G
Height:	13 ft	Type:	Rescue helicopter
Fuel:	100 galls	Unit cost:	$91,977
Empty weight:	4050 lb	Length:	57 ft 8 in
Crew:	2/4	Blade area:	115 sq ft
Top speed:	107 mph at 3800 ft	Oil:	8 galls
Cruise:	85 mph at 1500 ft	Gross weight:	5500 lb
Sea-level climb:	1240 ft per min	Hover ceiling:	3100 ft
Engine:	Pratt & Whitney Wasp Jr. R-985-AN-5	Range:	275 miles at 1500 ft
		Service ceiling:	14,800 ft

SERIAL INFORMATION

USCG No	Commissioned	Decommissioned
230	January 1947	December 1952 — crashed.
231	May 1947	September 1952 — crashed.
232	September 1946	
233	August 1946	
234	February 1949	September 1951 — crashed.
235	January 1950	
236	January 1950	May 1954 — crashed.
237	February 1950	
238	February 1950	

HO2S-1

US Navy BuNo 75689 ex-USAAF YR-5A 43-46613 to USCG Jan 1946. Crashed October 1950. While with USAAF modified to YR-5E.

US Navy BuNo 75690 ex-USAAF YR-5A 43-46618 to USCG January 1946, disposed of June 1946.

SIKORSKY HO4S-1G/2G/3G.

From the first flight in November 1949, until production ceased in March 1961, a total of 1281 Sikorsky S-55 helicopter variants were used by all the US military services, including the US Coast Guard, during the 1950s, playing an important and dramatic part in developing helicopter roles and techniques. First delivered to transportation and air rescue forces during the Korean conflict, the S-55 became the prime helicopter of many military air arms.

Of classic helicopter configuration, with a single main rotor and anti-torque tail rotor, the S-55 was the first of the Sikorsky helicopters with adequate cabin space and lifting ability to permit satisfactory operation in such roles as troop and transport and air/sea rescue. External sling loading, later to become a prominent feature of helicopter participation in power-line construction, was proved feasible by S-55 type helicopters in Korean hills as ammunition, food and medical supplies were airlifted between supply dumps and the front line.

On 28 April 1950, the US Navy placed a contract for an S-55 variant, the general purpose and anti-submarine observation HO4S-1 which was equivalent to the US Air Force and US Army H-19A helicopter. Deliveries of only ten built commenced on 27 December 1950, powered by a 550hp Pratt & Whitney R-1340 engine. The HO4S-2 was a version of the HO4S-1 for US Coast Guard use in the air/sea rescue role. Twenty-three of these more powerful Wright-engined models were acquired by the Coast Guard as HO4S-3Gs.

During November 1951 the first of seven Sikorsky HO4S-1G helicopters, USCG serial 1252, was delivered to the Brooklyn air station piloted by Lieutenant Commander Victor A Schmidt and Lieutenant Paul Porosky direct from the Sikorsky factory located at Bridgeport, Connecticut. Commander Theodore J Harris, Commanding Officer, greeted the new 'double-decked' helicopter, the newest addition to search and rescue (SAR) equipment at Brooklyn. Some Coast Guard documentation refers to the early S-55 helicopter as HO4S-2G, powered by the 550hp Pratt & Whitney R-1340 engines. Other USCG records indicate seven HO4S-1G and seven HO4S-2G were acquired.

In January 1952, a single Sikorsky HO4S-3G 1281 powered by a Wright engine was delivered to the Coast Guard. This was followed by a batch of thirteen HO4S-3G helicopters and a final batch of nine of the same model. Unit cost was $177,530. A large number of successful rescues were made with this helicopter, primarily using the hydraulic hoist and the Coast Guard-designed rescue basket. All these helicopters were fitted with 'Tug Bird', a Coast Guard design which provided the unique capability of towing boats and ships up to approximately 800 tons. Effective 1 July 1958, each USCG air station has at least one HO4S-3G helicopter equipped for towing vessels.

Eight helicopters of the HRS-3 version were borrowed from the US Navy and operated by the Coast Guard for several years. Continued development efforts resulted in provisions for all-weather capabilities, improved night illumination, extended range and payload plus flexibility and improvement of rescue devices. Under a US Department of Defense directive dated 6 July 1962, the designations allocated to US Navy variants were standardised with those of the US Air Force variants, and so the HO4S became the HH-19G.

Cruising speed was 80 knots, top speed 115 knots, and a range of nearly 400 miles. The HO4S-3G was fully equipped for instrument and night flying, having a hydraulic hoist to pick up personnel or equipment up to 400lb in weight. It was also equipped with padeyes for attaching slings to lift objects which would not fit in the cabin. On 23 January 1958, Sikorsky HO4S-3 1252 airlifted a 300lb foundation collar for new navigation radar equipment atop a tower at Elizabeth City USCG air station in North Carolina.

On 3 September 1959, a Coast Guard HO4S-3 1331 piloted by Lieutenant Benjamin Weems opened a landing field for emergency helicopter operations at a new heliport located at Napoleon Avenue and the river front in New Orleans, Louisiana. The land was donated by the New Orleans Board of Commissioners dock board.

A Sikorsky HO4S-3G helicopter 1298 from the USCG air station at Port Angeles, Washington state, participated in establishing an observation post on Blue Glacier, Mount Olympus, as part of the 1956 International Geophysical activities. The HO4S-3G flew in scientists and delicate instruments to an altitude of 7200 feet. Other supplies were dropped by US Air Force Douglas C-124 Globemaster transports with the Coast Guard helicopter providing rescue escort.

On 29 December 1954, a friend of the author, Boardman C Reed, flew in Sikorsky HO4S-3 1304 based at San Francisco air station, California, and was airlifted to Governers Island, Alameda. The Sikorsky c/n was 55.548 and date completed was 1951, the plate having the code 'HTCIH4 PCLH'. The engine was a 700hp Kaiser-Frazer built Wright R-1300-3 Cyclone 7.

Photo evidence shows 1253 as a HO4S-2, 1309 as a HO4S-3G marked 'SAN DIEGO' on the tail boom; 1255 as a HO4S-2G which was involved in a rescue net demonstration on 6 November 1956, at Oregon Inlet. On 15 May 1960, HO4S-3G 1252 was equipped with a loud hailer system demonstrated at Andrews AFB near Washington DC. During the winter of 1965 a Coast Guard HO4S-3 1308 was employed on ice reconnaissance flights up the Hudson River. It was piloted by Commander T F Epley, and based at Brooklyn USCG air station, New York. It was accompanied by the Coast Guard cutter *Sassafras* and while passing over the Tappan Zee area reported several vessels icebound and in need of assistance.

Sikorsky HO4S-3G '1304' of the USCG assists a small pleasure boat in distress off the coast of Miami, Florida, on 9 May 1963. Fourteen of the type were in service between 1951 and 1966. It was equipped with a 600-lb hydraulic hoist, was fully equipped for night flights, and was capable of towing boats. *(USCG)*

On 14 April 1956, a triangle played a vital part in saving the life of Seaman John Smith at the Public Health Hospital on Staten Island, New York, enabling Sikorsky HO4S-3 1301 to land in the grounds. Constructed of white stone, its three sides were thirty-two feet long with a large 'H' in the centre. Today most hospitals are equipped with a suitably marked landing aid area for use by Coast Guard helicopters.

Flotation gear was fitted to the Sikorsky HO4S helicopter and successfully demonstrated on HO4S-3 1308 at Elizabeth City air station on 5 December 1956. Most of the type were finished yellow overall, although the odd one had scarlet or red on the nose, tail plus centre section, including 1307 and 1309.

During October 1959, the two HO4S-3G helicopters based with the Coast Guard air detachment, located at West Bank, New Orleans, Louisiana, were flown out to Elizabeth City air station for overhaul. Over the past year they had flown 510 search and rescue (SAR) missions covering 53,943 miles representing 1,023 flight hours. This represented only a small part of their actual flying time as these helicopters performed many other tasks in cooperation with other Federal, state and local agencies, pilot training, logistic support plus sundry missions. After overhaul the two Sikorsky HO4S-3G were assigned to the US Coast Guard air station located at Traverse City, Michigan. The two replacement helicopters at New Orleans were two new Sikorsky HUS-1G helicopters 1332 and 1334.

TECHNICAL DATA

Manufacturer:	Sikorsky	Take-off power:	1 × 550 hp — 700 hp
Other designations:	HH-19G	Tail-rotor diameter:	8 ft 8 in
Rotor diameter:	53 ft	Designation:	HO4S-1G/2G/3G
Height:	13 ft 4 in	Type:	Rescue helicopter
Fuel:	190 galls	Unit cost:	$177,530
Empty weight:	5250 lb	Length:	42 ft 2 in
Crew:	2	Blade area:	2210 sq ft
Top speed:	112 mph	Gross weight:	7500 lb
Cruise:	91 mph	Passengers:	10
Sea-level climb:	1020 ft per min	Hover ceiling:	7900 ft
Engine:	Pratt & Whitney R-1340 (1G) (2G) Wright R-1300 (3G)	Range:	360 miles
		Service ceiling:	10,600 ft hover

SERIAL INFORMATION

1252 to 1258	HO4S-1G/2G
1281	HO4S-3G
1298 to 1310	HO4S-3G
1323 to 1331	HO4S-3G

SIKORSKY HO5S-1G

In conjunction with the US Coast Guard and the US Navy, the Sikorsky Aircraft Division of United Aircraft Corporation, of Stratford and Bridgeport, Connecticut, was kept active producing advanced models of previous helicopters. The first US helicopter to have metal rotor blades was the Sikorsky Model S-52 which was developed as a two-seat machine powered by a 178hp Franklin engine. It flew for the first time on 12 February 1947. It was later developed into the four-seat Model S-52-2 with a 245 hp Franklin 0-425-1 engine, and this variant was ordered by the US Marine Corps to replace the earlier Sikorsky HO3S. Deliveries commenced in March 1952. Procurement totalled seventy-nine which were designated HO5S-1.

Sikorsky HO5S-1 '1244' seen during a test flight at the Sikorsky factory in Connecticut, prior to delivery to the US Coast Guard during September 1952. Only eight of this type were acquired by the service covering two contracts. *(USCG)*

Under an initial contract Tcg 15531 the US Coast Guard ordered two HO5S-1G helicopters which were delivered in September 1952. A further six followed under contract Tcg 17701 being also delivered and commissioned by the Coast

Guard in September 1952. Sikorsky HO5S-1G 1246 crashed during March 1953. Unit cost was $82,928.

With a top speed of 90 knots, its small size, short range, and low life capability limited its effectiveness with the Coast Guard.

TECHNICAL DATA

Manufacturer:	Sikorsky	Engine:	Franklin 0-425-1 air cooled
Other designations:	S-52 H-18	Take-off power:	1 × 245 hp
Contract No.	Tcg. 15,531 & 17,701	Tail-rotor diameter:	6 ft 4 in two blade
Rotor diameter:	33 ft	Designation:	HO5S-1G
Height:	8 ft 8 in	Type:	Utility helicopter
Fuel:	55 galls	Unit cost:	$82,928
Empty weight:	2000 lb	Length:	29 ft
Crew:	2	Blade area:	43 sq ft
Top speed:	105 mph	Gross weight:	2770 lb
Cruise:	90 mph at 1,500 ft	Range:	190 miles at 1,500 ft
Sea-level climb:	250 ft per min	Service ceiling:	12,400 ft

SERIAL INFORMATION

USCG No	Commissioned	Decommissioned
1244	September 1952	Contract Tcg 15531
1245	September 1952	15531
1246	September 1952	Crashed March 1953. Contract Tcg 17701
1247	September 1952	17701
1248	September 1952	17701
1249	September 1952	17701
1250	September 1952	17701
1251	September 1952	17701

SIKORSKY HUS-1G

The Sikorsky S-58 was an all-purpose transport helicopter, first flown on 8 March 1954, and used in many roles by the US Navy and US Coast Guard, as well as the US Army and US Marine Corps. Many foreign operators both military and civil also operated the type. More than 1550 were manufactured by the Sikorsky Aircraft Division at Stratford, Connecticut. This helicopter had a seating capacity of a pilot and co-pilot, twelve to eighteen passengers or eight litters, or a net payload of 4000lb for a distance of 100 miles.

It had an alternative cargo capacity of 405 cubic feet, or 5000 lb capacity automatic touch-down release cargo sling to carry external loads and a 600lb capacity hydraulically-operated utility-hoist provision. Automatic stabilisation equipment was installed on versions produced for the US Navy, Coast Guard, Marines and Army. The four-bladed main rotor and four-bladed tail rotor were all-metal.

HISTORY

The Sikorsky HO4S helicopter operated by the US Navy and the US Coast Guard lacked range, the small usefull load severely restricted its usefulness in operations. However it must be said that the HO4S had great potential in the search and rescue (SAR) role with the Coast Guard. Sikorsky decided to design a similar but larger helicopter, the Model S-58. A prototype was ordered by the US Navy on 30 June 1952, with the designation XHSS-1.

Making its first flight on 8 March 1954, apart from its size, the new helicopter closely resembled the earlier S-55 (HO4S). The engine was a 1525hp Wright 1820 radial, located obliquely in the nose so that the transmission shaft ran at right angles to the engine straight into the gear box, beneath the rotor hub. Unlike the HO4S which had a nosewheel type landing gear, the new S-58 which was designated HUS-1G with the Coast Guard, had a tailwheel, and to facilitate shipboard stowage, the main rotor blades could be folded aft and the entire rear fuselage and tail rotor folded forward.

The US Navy placed production orders for the HSS-1 even before the first flight of the prototype. Subsequently, a transport and utility version was ordered by the US Marine Corps with the designation HUS-1 on 17 October 1954. These helicopters had the anti-submarine warfare (ASW) equipment deleted and provision for the carriage of up to twelve passengers in the cabin.

Six Sikorsky HUS-1G medium-range utility helicopters were acquired by the US Coast Guard during 1959 as a replacement for the less capable HO4S-1G which in 1962 became the HH-19G. The interior accommodation included side-by-side seating for pilot and co-pilot plus seating for ten passengers. It had a droppable external fuel tank and a range of 550 miles. The new HUS-1G helicopters were fitted with the latest electronic equipment and automatic stabilisation equipment (ASE) for navigation.

One of the Coast Guard bases operating the new HUS-1G was the air detachment at New Orleans in the Eighth Coast District. A helicopter was often the most advantageous means of searching the myriad bayou areas around the Mississippi River Delta. By means of the Coast Guard designed rescue basket and hoist, persons in almost any type of distress situations could be literally plucked to safety.

Sikorsky HUS-1G '1336' seen in flight over the skyline of New Orleans, Louisiana. Six of the type were delivered to the USCG during 1959, serving until 1962. The Air Detachment located at New Orleans in the 8th Coast Guard District, operated the type as a useful rescue tool in the Mississippi River Delta. *(USCG)*

It is unfortunate that history records that the Coast Guard had bad luck with the operation of the HUS-1G. Two were lost in Tampa Bay, Florida, within an hour of attempting to go to the aid of a US Air Force Boeing B-47 Stratojet which had ditched. Another was lost at sea in the Gulf of Mexico after the helicopter struck the rigging of a fishing boat at night during the hoisting operation of an injured crewman.

Under a US Department of Defense directive dated 6 July 1962, the designation allocated to US Navy and US Coast Guard types were standardised with those of the US Air Force. The HUS-1G became the HH-34F. The USCG serial numbers allocated were '1332 to 1336' and 1343'.

The Coast Guard elected to abandon their plans to purchase more Sikorsky HUS-1G (HH-34F) helicopters as, by the time, the new Sikorsky Model S-62 was on the drawing board, this becoming the HH-52A Seaguard with the Coast Guard.

TECHNICAL DATA

Manufacturer:	Sikorsky	Take-off power:	1 × 1525 hp
Other designations:	HH-34F, S-58	Main-rotor diameter:	56 ft
Rotor diameter:	56 ft	Designation:	HUS-1G
Height:	14 ft 4 in	Type:	Utility/rescue helicopter
Fuel:	198 galls	Length:	65 ft 0 in
Empty weight:	7900 lb	Blade area:	2460 sq ft
Crew:	2	Gross weight:	13,000 lb
Top speed:	122 mph	Passengers:	12
Cruise:	97 mph	Hover ceiling:	5500 ft
Sea-level climb:	1100 ft per min	Range:	247 miles
Engine:	Wright R-1820-84	Service ceiling:	9500 ft hover

SIKORSKY HH-52A SEAGUARD

The US Coast Guard selected a version of the commercial Sikorsky Model S-62 during 1962 as a replacement for its Sikorsky HH-34F (HUS-1G) helicopter in the utility and search and rescue (SAR) role. Deliveries to the service commenced during January 1963, and ninety-nine helicopters were purchased off-the-shelf from the Sikorsky Aircraft Division of United Technologies Corporation of Stratford, Connecticut. The new helicopters were modified and flight tested for Coast Guard use. Although unique in not being acquired by any of the other US military services, it did initially have the US Navy designation HU2S-1G, but under a US Department of Defense directive dated 6 July 1962, all designations allocated to US Navy aircraft were standardised with those of the US Air Force. So the new helicopter was designated HH-52A and given the name Seaguard.

A unique design, the single turbine-powered Sikorsky S-62, the first amphibious helicopter built with a flying boat type hull, was selected by the US Coast Guard as its medium-range search and rescue (SAR) helicopter. It was the first American turbine-powered helicopter to be certificated under the FAA's revised and more stringent helicopter transport regulations. The FAA certificate permitted the S-62 to carry a pilot, co-pilot and up to eleven passengers. It was capable of operating from land, water, ice, snow, swamp and almost any other surface.

The first flight of the S-62 took place on 22 May 1958, and the manufacturer claimed it was the first helicopter in history to fly from datum with proven mechanical components having extended overhaul periods. It became the first jet-powered helicopter in the US Coast Guard inventory, and this amphibian attained the services long standing requirement, dating back to the birth of Coast Guard Aviation, for a Flying Life Boat. It was powered by a General Electric T58-GE-8B turbine engine, cruised at 85/90 knots, having a maximum range of over 400 nautical miles. The design gross weight was 7900lb with a maximum gross weight of 8300lb. In addition to shore base operations from the USCG air stations, the HH-52A was deployed on Coast Guard icebreakers and cutters equipped with a flight deck on the stern. The purchase price for the aircraft ranged from approximately $250,000 to $500,000. The last of the ninety-nine Seaguard helicopters was delivered during 1969.

Several special features for the rescue role included a folding platform which could be dropped down to aid survivors from the water, and a winch position above the door. One of its claims to fame is that it has rescued more persons in distress than any other helicopter in the world.

The contract for the Sikorsky HH-52A was settled on 21 June 1962, and a release CPI-06-21-62-G depicted an artist's impression of s Sikorsky HU2S-1G carrying USCG serial 9028'. A set of six Coast Guard photos CPI-020662(A) dated 6 February 1962,

Sikorsky HH-52A '1356' seen en-route from the factory to USCG Air Station Salem, Massachusetts, on 25 January 1963 on delivery. It was the fourth production aircraft from ninety-nine acquired by the service. Being an amphibian it soon proved its value and versatility. *(Sikorsky)*

depicted a HU2S-1G with the US civil registration N880 in full US Coast Guard livery. It performed well in eight to ten foot waves and at that time the Coast Guard were negotiating to purchase four of the new helicopters. It had a carrying capacity of 2900lb. It was equipped with automatic stabilisation equipment, able to fly approximately 190 miles from base, pick up an injured person with the rescue hoist or by landing on the water, and return to home base with a ten per cent margin of fuel remaining.

On 9 January 1963, Admiral E J Roland, Commandant of the US Coast Guard, accepted the first four HH-52A Seaguard helicopters 1352, 1353, 1354 and 1355 at the Sikorsky plant located at Stratford, Connecticut. On Friday 25 January 1963, Commander J M Waters USCG, Commanding Officer of the the USCG air station at Salem, Massachusetts, took delivery of the first HH-52A Seaguard helicopter 1352 at Stratford and flew it back to Salem air station, one of three to initially equip the unit.

The new HH-52A soon proved its value and versatility, from the day it entered service at Salem as the first turbine-powered, boat-hulled helicopter in the history of Coast Guard Aviation. When hurricane 'Betsy' struck the New Orleans area in the autumn of 1965 Coast Guard pilots flew the HH-52A around the clock. With a constant threat from power lines, trees, structures and water, and with the wind and flood still strong, they rescued nearly 1200 people from a wide area that was devestated by the hurricane.

During January 1967 a Miami air station based HH-52A 1384 flew a US Public Health Service doctor to the Coast Guard cutter *Halfmoon* to tend to a fisherman who was wounded during a shooting spree approximately seventy-six miles east-south-east of Marathon, Florida. The cutter had earlier seized four fishing vessels and their crews and turned them over to the Federal Bureau of Investigation (FBI) at Marathon.

VERSATILITY

The Coast Guard helicopter will continue to be a primary rescue tool into the foreseeable future. During 1980 over 100,000 refugees fled from Communist-ruled Cuba. Many risked their lives in unsafe craft to cross the Straits of Florida. The rescue of those on board the *Olo Yumi* is just one example of many situations confronting the Coast Guard. On the morning of 17 May, 1980, the pleasure craft *Olo Yumi,* carrying fifty-two persons, sank when the people on board panicked because of rough seas, ran to the stern, and caused water to pour over the transom. A Sikorsky HH-52A Seaguard on patrol from the USCG cutter *Courageous* (WEMEC-622) sighted the refugees in the water and commenced rescue operations. Eleven survivors were winched into the helicopter. Other Coast Guard helicopters and the cutter *Courageous* rescued thirty-eight survivors and recovered ten bodies. The pleasure boat had been grossly overloaded.

Not long after introduction into Coast Guard service the HH-52A joined the USCG ice operation detachment on board powerful icebreakers such as the *Burton Island,* and the 399-ft *Polar Star,* which operated in Antarctica. The Coast Guard and its helicopters were involved in the Antarctic Operation Deep Freeze in 1973 on the icebreaker *Glacier.* The USCG Ice Operations programme still covers a wide spectrum of icebreaking services which assist in moving maritime traffic, preventing and controlling flooding resulting from ice accumalation on domestic waterways, and supporting scientific research and other US national interests in the polar regions.

The Coast Guard HH-52A helicopter played a major role in many operations. In its ice reconnaissance mission it flew ahead of the ships searching for openings — called 'leads' — in the ice. Once found, the job of breaking ice begins. The helicopter also makes cargo deliveries to the numerous scientific outposts dotting the Arctic and Antarctic whiteness.

Coast Guard aviation crewmen serviced the Sikorsky Seaguard helicopter on board ship, a demanding and life supporting task. they maintained the survival gear of the helicopter detachment such as the survival vests, rafts and aux flotation gear on the HH-52A. The aviation electricians maintained all of the electrical systems on the helicopter including the automatic stabilisation and compass systems.

On the Tuesday before Christmas, 22 December 1964, Sikorsky HH-52A Seaguard 1382 guided by radar from the airport and Coast Guard base 217ft above sea level at Arcata, in Humboldt County, California, plucked twenty persons from the storm ravaged and flooded countryside and flew them to safety. With a crew of three plus three more survivors the helicopter took off again when it disappeared from the radar screens. It was four days later before a US Navy helicopter pilot was able to reach the crash site in rugged country twenty miles north of Eureka. The three crewmen and three civilians on board were dead. The pilot was Lieutenant Commander Donald L Prince USCG and the co-pilot was from the Royal Canadian Navy — Sub Lieutenant Allen E Alltree — working with the US Coast Guard in an exchange programme. The crewman was Aviation Electricians Mate James E Nininger Jr USCG. Today a plaque on a stone plinth records the tragedy. 'They gave their lives on their mission.'

On 28 January 1968, a fifteen-foot sailboat with seven Cuban refugees on board was spotted by the Liberian registered freighter *Arcturus* ten miles east of Elliot Key, off Florida.One of the refugees, Jesus Palalz, suffered with an infected foot and was airlifted from a USCG 40-ft patrol boat by a HH-52A helicopter from the Miami air station and flown to the Mercy Hospital in Miami.

A radio message from the Bremen-based German freighter *Clivia* position 270 miles north-east of Miami, Florida, requested help for a 74-year old Canadian passenger who was suffering from an inflamed eye and was unconscious. Date was 31 May 1966. A HH-52A Seaguard helicopter from the USCG air station Miami located at the Opa Locka airport was alerted but due to the distance involved had to refuel from a USCG Grumman HU-16E Albatross amphibian at Marsh Harbor, Great Abaco Island. After refuelling the helicopter reached the freighter then 100 miles north-east of Marsh Harbor. Hovering over the stern of the *Clivia* the patient was airlifted to a hospital in Nassau.

Piloted by Lieutenant W H Goldhammer USCG, a HH-52A flew out to the US Navy minesweeper *Turkey* during June 1964 to hoist aboard the sick Seaman Apprentice Raymond Beaudry and airlift him to hospital on Staten Island, New York. The minesweeper was located thirty miles off Asbury Park, New Jersey.

On 31 July 1976, a distress call was received by the Coast Guard air station at Elizabeth City, North Carolina, referance a critically ill seaman aboard a drilling platform which had departed Galveston, Texas, under tow by a tug, and en route to Halifax, Nova Scotia. Position was given for the rig *Canmor Explorer II* as thirty-five miles southeast of Cape Hatteras. Initially a USCG Lockheed HC-130 Hercules from Elizabeth City located the tug and drilling platform and then directed HH-52A Seaguard 1400 to the platform where it landed on the helicopter deck to pick up the injured seaman.

With the introduction of the new Coast Guard 210-ft cutter class equipped with a helicopter flight deck, a new concept in the Coast Guard task or mission came into being. Trained crewmen positioned in the safety netting around the flight deck had the HH-52A helicopter secure within four seconds of touch-down. Four quick-release webley straps, fitted with hooks, lashed down the helicopter. After engine shut-down the operation was completed with fourteen quick-release chain lashings each designed with a strength of 20,000lb. The cutter *Reliance* (WPC-615) was the first USCG vessel to operate in the Gulf of Mexico with a helicopter pad, trials taking place with a HH-52A during October 1964, just four months after the cutter was commissioned. It was home-based at Corpus Christi, Texas.

COORDINATION

On Wednesday 5 February, 1964, the 12th US Coast Guard District Rescue Coordination Center in San Francisco, California, received a distress call from a 105-ft chartered yacht *Hattie D*, a converted World War II US Navy vessel. The pleasure voyage of the yacht had began in Seattle, Washington, on 24 January and the distress call indicated a position approximately 20 miles south of Cape Meendocino, California. The vessel had lost its rudder in collision with a floating object during a storm.

Coast Guard rescue units were dispatched immediately to assist the yacht. Aircraft from the USCG air station at San Francisco plus the 82-ft patrol cutter *Point Ledge* (WPB-82334) from Noyo, and the cutter *Avoyal* from Fields Landing were soon manned and on their way to the scene. An attempt was made to airdrop a pump to the yacht but the strong gale force winds carried it away. With the arrival of a Sikorsky HH-52A helicopter, a Lockheed HC-130B Hercules and a Grumman HU-16E Albatross they alternated in communicating the position of the yacht to the USCG Rescue Coordination Center.

In two trips the HH-52A helicopter airlifted ten men, one woman and a German shepherd puppy from the stern of the sinking yacht *Hattie D.* In the first lift it took six persons and deposited them on a beach, one and a half miles north of Loleta on Copenhagen Road, leaving them in the care of the Humboldt County sherif. On return to the sinking yacht the HH-52A picked up the remaining five persons, then flew the survivors to Arcata Airport. Some fifty minutes after rescue had been completed the *Hattie D* sank.

On 7 November 1964, smoke from a burning vessel was sighted by Coast Guard personnel from the window of the Merrimac Rover Station, Newburyport, Massachusetts. The 65ft training ship *Aquanaut II* had completed a charter mission and was en route south when a blaze erupted from an oil burner in the engine room. Immediately two 30-ft USCG utility boats from

Seen flying over the picturesque San Francisco Bay area is HH-52A Seaguard '1384' based at the local USCG Air Station. This was the first turbine powered boat-hulled helicopter to serve the USCG. The type was finally retired in 1989, its claim to fame being its rescue record of more persons in distress than any other helicopter in the world. *(USCG)*

the Merrimac River Station and from Hampton Beach, New Hampshire, went to the scene to commence combating the fire. By this time the burning vessel was positioned two miles east of Merrimac Jetty, Plum Island, Massachusetts. A Sikorsky HH-52A Seaguard helicopter 1362 was called in from Salem USCG air station to carry foam to enable the two utility boats to continue to fight the fire. The training ship was owned by the US National Youth Science Foundation. Weather conditions at the time of the fire were good: wind eighteen miles per hour, temperature 49 degrees, weather clear and a four-foot sea.

During 1989 the last Sikorsky HH-52A Seaguard helicopter was retired from US Coast Guard service, so ending a thirty-year association. Many are preserved in museums throughout the United States, in pleasure parks and schools as a lasting tribute to a Coast Guard helicopter with yet another fine record of life saving in the history books of Coast Guard Aviation.

TECHNICAL DATA

Manufacturer:	Sikorsky	Tail-rotor diameter:	8 ft 9 in
Other designations:	HU2S-1G	Designation:	HH-52A Seaguard
Rotor diameter:	53 ft 6 in	Type:	All-weather rescue helicopter
Height:	14 ft 2 in		
Fuel:	uses 63 galls per hour	Unit cost:	Airframe $¼-½ million
Empty weight:	4903 lb	Length:	44 ft 7 in
Crew:	1/2 pilots + 1 crewman	Gross weight:	8300 lb
Top speed:	109 knots	Passengers:	11
Cruise:	85/90 knots (98 mph)	Hover ceiling:	1700 ft
Sea-level climb:	1070 ft per min	Range:	400 plus nautical miles = 474
Engine:	G.E. T58-G.E. 8 turbo shaft		
Take-off power:	1 × 845 shp		

SERIAL INFORMATION

Serial blocks:
1352 to 1379 = 28
1382 to 1413 = 32
1415 to 1429 = 15
1439 to 1450 = 12
1455 to 1466 = 12
Total = 99

Sikorsky constructors numbers
1370 c/n 62-048
1406 62-091
1411 62-096 sold as TF-GNA to the Icelandic Coast Guard
1418 62-102
1443 62-126

SIKORSKY HH-3F PELICAN

During 1968 the US Coast Guard procured the first of forty Sikorsky HH-3F Pelican helicopters for the Medium Range Recovery (MMR) mission, and thirty-six remain in service today. Some twelve are located at the Clearwater, Florida, air station, the largest establishment in the service in respect of personnel and aircraft. The air station is heavily involved in the drug interdiction conflict. This twin-engined, medium-range, amphibious, all-weather helicopter has extended the off-shore search and rescue (SAR) and patrol capabilities of the US Coast Guard.

The type is based on the Sikorsky Model S-61B helicopter, with the engines and airframe being the same as the US Air Force CH-3C, however the avionics, fuel system and instrumentation are different. It has a maximum speed of 142 knots with a normal cruise of 120 knots. The HH-3F can fly 285 nautical miles (528 km) out to sea, hover for twenty minutes, rescue six persons and return to land at its base. It carries two pilots, navigator and crew chief, with seats for six passengers. It also has provision for twenty passengers or nine litters. With the HH-3F on the water, a special rescue platform can be extended from the cabin door for retrieval of disabled survivors. In addition to SAR assignments, the Pelican missions include aids to navigation, port security, law enforcement, inter-agency cooperation in natural disasters, military readiness and logistics.

The US Coast Guard purchased forty HH-3F Pelicans, the last one being delivered during 1973. The price quoted for the airframes, not including the two engines, avionics and furnished equipment, was approximately $900,000. The search and weather radar is fitted in a nose radome offset to port, this being a distinctive feature of the HH-3F. The fuselage incorporates a loading ramp, and naturally the fuselage is watertight to permit it to alight on water.

HISTORY

Initially developed for the US Navy, the Sikorsky S-61 helicopter, was contracted on 24 december 1957. This new helicopter was considerably larger than previous models having the ability to carry both search equipment including dipping sonar, plus 84 lb of weapons, as well as instrumentation for all-weather operations. A watertight hull design was used since most of the operation life would be spent over water, and stabilising floats provided a convenient housing for the retractable wheels, making the S-61 a true amphibian.

Power was provided by two General Electric T58 turboshaft engines side by side above the main fuselage located close to the gearbox driving the five-blade main rotor. Primary search equipment was the Bendix AQS-10 or AQS-13 sonar, with a coupler to hold hovering altitude automatically in conjunction with a Ryan APN-310 Doppler and radar altimeter. A Hamilton Standard autostabilisation system was incorporated for all-weather capability.

The initial batch of seven trials aircraft carried the designation YHSS-2. Under a US Department of Defense directive dated 6 July 1962, the designation allocated to US Navy variants were standardised with those of the US Air Force. In this unified system the Sea King was subsequently redesignated SH-3. First flight was made on 11 March 1959, and US Navy acceptance trials with the Navy Board of Inspection and Survey began during February 1961, with delivery to fleet units commencing in September 1961.

OPERATIONS

During 1990 the Sikorsky HH-3F Pelican helicopters of the US Coast Guard were heavily involved in the OPBAT Operation — Bahamas, Turks and Caicos mission, involved in the apprehension of drug traffic. Nightsun, a one million candle power lamp is strapped to the helicopter. Pelican 1491 crashed in Mexico during November 1988, but was recovered, repaired and placed back in USCG service at Mobile air station in Alabama. Service use of the HH-3F Pelican will be supplemented and possibly finally replaced by the Sikorsky HH-60J Jayhawk helicopter later in 1990s. The HH-3F has a hover in-flight refuelling capability (HIFR) from surface vessels.

The many rescue operation involving the giant Pelican helicopter would fill a huge volume. The following have been taken at random from the files.

On 24 March 1971, a Sikorsky HH-3F from Brooklyn air station, New York, was called out after the forty-two foot sloop *Big Schott* had requested assistance off the coast of Cape May, New Jersey. Fifty-five mile an hour winds had torn the mainsail making the sailboat disabled. Hours later the *Cape Gull,* a ninety-five foot patrol boat from Atlantic City, New Jersey, arrived on the scene and took the *Big Schott* in tow to Cape May. The four crewmen were uninjured and the sloop sustained no damage.

Seaman Apprentice Paul D Smith, aged 17, who suffered head injuries in a fall aboard ship on 19 April 1972, was initially rescued by the USCG cutter *Klamath* some 300 miles southwest of San Francisco. The following day a HH-3F Pelican helicopter '1435' picked up the injured seaman from the cutter, returning him to the San Francisco air station, where he was taken by ambulance to the US Public Health Service Hospital in San Francisco for treatment.

During July 1979 a Sikorsky HH-3F 1473 was involved in the rescue of a boating accident victim aboard the forty-foot *Sport Fisherman* off Cape Cod, New Jersey. The victim was airlifted to a local hospital for treatment. The HH-3F was from Cape Cod air station.

During October 1980 as the Holland Cruise ship *Prinsedam* steamed through the Gulf of Alaska on a scheduled month-long voyage from Vancouver, Canada, to Singapore, a fire broke out. The huge ship was abandoned by its 320 passengers and approximately 200 crew members as the fire roared out of control. All of the passengers clambered into lifeboats, and three US Coast Guard cutters and the 1000-foot US *Supertanker* went to the aid of the stricken ship. Firefighters and some of the crew were airlifted to safety by Sikorsky HH-3F helicopters of the US Coast Guard from Kodiak air station. This rescue included

Of the forty Sikorsky H-3F Pelicans acquired from the USCG thirty-six remained in service by 1991. They were heavily involved in the apprehension of drug traffic in the Caribbean area. Depicted is HH-3F '1434' on a patrol flight over the US seaboard. It will be replaced by the HH-60J. *(Sikorsky)*

HH-3F 1475 from Kodiak. Eight days later the *Prinsedam* sank.

On Monday 16 March 1987, the Soviet cargo ship *Komsomolets Kirgizii* reported to be carrying a cargo of flour from Nova Scotia to Cuba, began to capsize in heavy seas 210 miles off

New Jersey. It sent out distress signals after the vessel's engines failed, and it was soon listing to port after a broadside battering from huge waves for two days. All thirty-seven people on board including a woman and a baby were airlifted from the deck of the Soviet ship by Sikorsky HH-3F helicopters of the US Coast Guard. The rescue involved coordination between two USCG air stations, three USCG districts, three HH-3F helicopters and two Lockheed HC-130 Hercules aircraft. Only one crew member was slightly injured with a cut finger, and all thirty-seven people were taken to a motel in Philadelphia before flying home to Moscow from New York. The Soviet crew were able to meet and thank the Coast Guard the following day at a reception held at the White House.

Further rescue exploits can be found in *A History of US Coast Guard Aviation,* also published by Airlife.

TECHNICAL DATA

Manufacturer:	Sikorsky	Designation:	HH-3F Pelican
Other designations:	HH-3E, HH-3C	Type:	All-weather SAR helicopter
Rotor diameter:	62 ft	Unit cost:	$900,000 airframe only
Height:	18 ft 1 in	Length:	57 ft 3 in
Fuel:	1105 galls	Blade area:	445 sq ft
Empty weight:	13,422 lb	Oil:	7 galls
Crew:	4	Gross weight:	22,050 lb
Top speed:	136 knots	Passengers:	20
Cruise:	130 knots	Hover ceiling:	6200 ft
Sea-level climb:	1240 ft per min	Range:	300 nautical miles
Engine:	G.E. T58-G.E.-5	Service ceiling:	10,500 ft
Take-off power:	2 × 1500 shp		
Tail-rotor diameter:	10 ft 4 in		
Serial information:			
1430-1438 = 9			
1467-1497 = 31			

SIKORSKY CH-3E

During 1989 the US Coast Guard acquired a handful of Sikorsky CH-3E rescue helicopters from the US Air Force of which all but two underwent rework and modification at the Elizabeth City facility located at the air station in North Carolina. These two were retained as replacements and for spares. The CH-3E carries the name Sea King with the US Navy, and Jolly Green Giant with the US Air Force. They are to supplement the aging Sikorsky HH-3F Pelican helicopters in use by the service. Modification by the Coast Guard includes the addition of an auxiliary fuel tank for increased range on Search & Rescue (SAR) missions, a Loran 'C' navigation system and the fitting of an APN-215 radar in the nose.

The Sikorsky S-61 helicopter had originally been developed for the US Navy. During April 1962, the US Air Force showed an interest and borrowed three US Navy HSS-2s to fly re-supply missions to radar outposts sited in the Atlantic. These were known as Texas Towers. Three more were obtained later in 1962, the six in service with the US Air Force then becoming CH-3Bs.

In order to meet their requirements for a long-range transport helicopter, the US Air Force ordered the same basic design in

Sikorsky CH-3E '9691' ex US Air Force, seen after rework for the USCG in the hangar at Elizabeth City, North Carolina. This helicopter went into service during 1990 at CGAS Traverse City, Michigan. The USCG received a handful of CH-3E helicopters from the USAF during 1989, retaining some at Elizabeth City. *(USCG)*

November 1962, this becoming the CH-3C. It had the same powerplant, rotors and transmission as the basic Sikorsky S-61, but featured a new rear fuselage design incorporating a loading ramp for vehicles. The first flight was made on 17 June 1963, and the following month it was the winner of a competition for a long-range rotary-wing support system.

The substitution of 1500 shp T58-GE-5 engines for the 1300 shp GE-1s in the CH-3C changes the designation to CH-3E in February 1966. Forty-five were built with the new engines and forty-one early CH-3Cs were updated.

First to appear in US Coast Guard livery after rework and modification at Elizabeth was 9691 ex-USAF 63-9691, the second being 2793 ex-USAF 65-12793, it being coincidental that both had previously served with the 89th Military Airlift Wing whilst with the US Air Force. A third CH-3C ex-USAF 62-12578 destined for USCG service had also served with the 89th MAW. The unconverted helicopters wer believed to be ex-USAF 65-12788 and 62-12789.

TECHNICAL DATA

Manufacturer:	Sikorsky	Tail-rotor diameter:	10 ft 4 in
Other designations:	S-61R	Designation:	CH-3E
Rotor diameter:	62 ft	Type:	Rescue helicopter
Height:	18 ft 1 in	Length:	73 ft
Fuel:	700 galls	Blade area:	3019 sq ft
Empty weight:	12,423 lb	Oil:	7 galls
Crew:	2/3	Gross weight:	22,050 lb
Top speed:	164 mph	Passengers:	25
Cruise:	154 mph	Hover ceiling:	14,700 ft
Sea-level climb:	1520 ft per min	Range:	760 miles
Engine:	2 × 1500 shp GE T58-GE-5	Service ceiling:	13,600 ft

SERIAL INFORMATION

USCG No		Decommissioned
9691	ex-USAF 63-9691	89th Military Airlift Wing
2793	ex-USAF 65-12793	89th Military Airlift Wing
	ex-USAF 62-12578	89th Military Airlift Wing
	ex-USAF 63-9679	see below
	ex-USAF 65-12788	
	ex-USAF 65-12789	
	ex-USAF 64-14234	returned to Davis Monthan, Arizona April 1990
	ex-USAF 65-5697	returned to Davis Monthan, Arizona April 1990
	ex-USAF 63-9679	returned to Davis Monthan, Arizona April 1990

9691	ex-USAF 63-9691	CGAS Traverse City, Michigan.
2791	ex-USAF 62-12791	CGAS Traverse City, Michigan.
2793	ex-USAF 62-12793	CGAS Traverse City, Michigan.
2578	ex-USAF 62-12578	ARSC Elizabeth City, North Carolina.
2788	ex-USAF 65-12788	ARSC Elizabeth City, North Carolina.

SIKORSKY HH-60J JAYHAWK

During the last month of 1990 the new Sikorsky HH-60J Jayhawk, newest addition to the helicopter fleet of the US Coast Guard, entered operational service with the Aviation Training Center (ATC) at Mobile, Alabama. A total of thirty-two of this Medium Range Recovery (MRR) were initially contracted, with pilot and crew transition training and evaluation procedures involving this new asset to the service. The US Navy Air Systems Command served as the procuring agent for what was termed as the 'three-two-one' buy. Three missions, two configurations and one contract.

The genesis of the Jayhawk programme can be found in a joint US Navy-Coast Guard procurement linking the HH-60J order with that of eighteen HH-60H strike rescue-special warfare support aircraft for the US Naval Air Reserve. Both helicopters are direct derivatives of SH-60F inner-zone anti-submarine warfare helicopter.

Appearance of the HH-60J on the US Coast Guard flight line is the continuation of a long tradition. During more than forty-five years of association a dozen Sikorsky helicopter types have served with the USCG, more than two hundred and fifty machines in all. The Jayhawk is replacing the venerable Sikorsky HH-3F Pelican in the MRR mission, which has been in the USCG inventory since 1960. In its search and rescue (SAR) role, the new helicopter can fly up to three hundred miles offshore and maintain an on-scene endurance of forty-five minutes. With a crew of four, made up of pilot, co-pilot, flight mechanic, plus an additional aircrewman or rescue swimmer, the HH-60J can transport at least six survivors from the maximum SAR radius.

The HH-3F Pelican helicopters will reach the end of their economic service by the end of the decade. As mentioned earlier the new HH-60J is a close derivative of the Sikorsky SH-60F Sea Hawk for the US Navy and builds on the $1.3 billion invested in the development of the H-60 helicopter series. It combines the best of combat attributes, seaworthiness, integrated systems, support ability and helicopter performance to provide the Coast Guard with the optimum MRR solution.

Entering service with the US Army as the UH-60A Black Hawk, the Sikorsky S-70A helicopter was selected by the US Navy during September 1977, becoming the SH-60B. It made its first flight on 11 February 1963. A second variant of the SH-60B Sea Hawk entered full-scal development for the US Navy in March 1985, when the SH-60F was selected as an inner-zone ASW — anti-submarine warfare — helicopter to replace the SH-3H Sea King.

A combat search and rescue (SAR) special warfare support version of the Sea Hawk was ordered by the US Navy during September 1986, designated HH-60H and called Rescue Hawk. A crew of four were provided with a capability to rescue eight survivors. First flown on 17 August 1988, the HH-60H entered service during August 1989, replacing the Sikorsky HH-3A Sea King.

With its proven airframe derived from the US Navy H-60 family, the extended range, advanced avionics, communication suites, including a search/weather radar, the Jayhawk offers a dramatic upgrade in helicopter capability in US Coast Guard drug interdiction missions. It can detect and interdict smugglers both on the surface and in the air. With its night vision goggles (NVG) capability, secure communications, data link, high speed and ample range, it can cover more area for an extended time without being detected visually or electronically.

It can provide logistics support for aids to navigation or rescue operation. Tasks such as airlifting a one-ton generator to a remote lighthouse, or rushing a dewatering pump to a sinking vessel are all in a day or night work for the Jayhawk, with its unsurpassed internal and external lift capacity.

Requirements call for the helicopter to perform in violent storm-force winds up to sixty-three knots (Beaufort Scale 11) and exceptionally heavy sea states. With its advanced avionics including RDR-1300 search-weather radar, the helicopter offers a dramatic upgrade of ship-helicopter team capability in drug interdiction misions. It is fully compatable with the helicopter decks of the 378-ft Hamilton class, and the new two hundred and seventy foot Bear class Coast Guard cutters.

From approach procedures to power control in the HH-60J there is very little difference from the procedures taught in the HH-3F. The systems, though naturally more advanced in technology, are based on the same principle of operation as both the USCG HH-3F and HH-52A Seaguard. Initial courses at both the USCG Air Training Center and the Aviation Technical Training Center (ATTC), the latter at Elizabeth City, on the the new helicopter were highly successful.

The HH-60H helicopter is a single main rotor, twin-engined aircraft. It is configured with a canted tail rotor, controllable stabilator, conventional fixed landing gear, emergency flotation, external cargo hook of six thousand pound capability, a six hundred pound rescue hoist, and pylons for carrying external stores such as fuel tanks. In addition it is equipped with a flight-rated auxilary power unit (APU), an anti-ice system, a fire-extinguishing system, plus environmental control, automatic flight control, a single-point pressure refuelling, helicopter in-flight refuelling system and the necessary avionics and instrumentation for instrument flight and mission accomplishment.

Two General Electric T700-GE-401C front-drive turboshaft engines power the HH-60J Jayhawk. These engines incorporate an integral inlet particle separator and modular construction. Under standard conditions, sea-level and fifty-nine degrees Fahrenheit, the engine has a maximum continuous shaft horsepower rating of sixteen sixty-two, and intermediate shp rating of eighteen hundred for thirty minute duration, and a selectable contingency power rating of nineteen forty shp available for two and a half minutes duration.

Primary missions for the HH-60J are search and rescue (SAR) plus Enforcement of Laws & Treaties (ELT) patrol. Secondary missions include support of US Navy missions as maybe required by Maritime Defence Zone (MDZ) operations. The helicopter also supports other Coast Guard missions as needed. It conforms to all contractural performance demands.

The general arrangement of the exterior of the aircraft for Coast Guard specifications include the left inboard, left outboard, and right pylons designed to accommodate BRU-14/A weapon-stores racks. The pylons have wiring and tubing provisions for three one hundred and twenty gallon auxiliary fuel tanks. Emergency flotation bags are installed on the stub wing fairing of the main landing gear on both sides of the helicopter.

The avionics/navigation package uses a fifteen fifty-three data bus to integrate almost all aircraft avionics functions. A control display unit CRT and keyboard is available to each pilot and the crew member to provide access and display of communication, navigation, and fuel systems. The co-pilot also has a display control panel keyboard and joystick for control and display of the tactical navigation scenario.

Upgrades for the Jayhawk were planned by Sikorsky via Engineering Change Proposals (ECP) with the goal of minimum retrofit. Other less critical upgrades were addressed via Block Upgrade programmes. One candidate for upgrading included fuel boost pumps. Certain combinations of pressure altitude and temperature precluded the use of JP4 grade fuel. The addition of fuel boost pumps provides the capability to use both JP4 and JP5 fuel in all operating conditions. Production line constraints required retrofit of the first five HH-60J Jayhawks '6001/2/3/4/5.'

A further important upgrade involved the HF (high frequency) antenna modification, with the existing HF longwire antenna being replaced with a tuned monopole antenna similar to that on the USCG Aérospatiale HH-65A Dolphin installation. The one-million candle power 'Nightsun' high intensity searchlight was modified so it could be mounted on the starboard external stores pylon. The Coast Guard has adopted an excellent policy on avionic modifications and enhancements to minimise development costs and facilitate better management and configuration control. The block upgrade cycle for the HH-60J is four years, commencing with Block Upgrade I set for October 1991 during Fiscal Year 1992. Additional planned upgrades include an improved aircraft emergency floation system and a forward looking infra-red radar (FLIR) system.

The US Coast Guard requested authority via the US Department of Transportation, to procure fifteen additional HH-60J helicopters specifially for the OPBAT Operations in the Bahamas/Turks and

The first production Sikorsky HH-60J Jayhawk '6001' ex US Navy BuNo 163801 seen during a test flight from the Stratford, Connecticut, factory. It is a derivative of the SH-60F Seahawk developed for the US Navy. It will initially supplement and later replace the HH-3F Pelican in USCG service. *(Sikorsky)*

Caicos Islands for support in the apprehension of drug traffic. Funding for these additional helicopters commenced in Fiscal Year 1991, continuing through Fiscal Year 1994, with projected deliveries commencing in January 1994. The first production contract was signed between the US Coast Guard and Sikorsky on 29 September 1986, the acquisition being for twenty-four helicopters initially, with a lot option for eleven more.

First flight of the Sikorsky HH-60J Jayhawk was on 8 August 1989, with the official roll-out to the Coast Guard on 14 September 1989, which coincided fifty years to the day with the first successful helicopter flight of aviation pioneer Igor I Sikorsky. The last of thirty-five Coast Guard HH-60Js will be completed in 1993.

TECHNICAL DATA

Manufacturer:	Sikorsky	Main Rotor diameter:	53 ft 8 in Four blades
Other designations:	H-60 S-70B	Tail Rotor diameter:	11 ft four blades
Rotor diameter:	53·66 ft	Designation:	HH-60J Jayhawk
Height:	17·16 ft	Type:	Medium range recovery helicopter
Fuel:	590 galls (US)	Length:	66·85 ft
Empty weight:	13,417 lb	Blade area:	2261 sq ft
Crew:	4	Oil:	5 galls
Top speed:	145 mph at 5,000 ft	Gross weight:	21,246 lb
Cruise:	146 knots	Passengers:	6
Sea-level climb:	576 ft per min	Hover ceiling:	5000 ft
Engine:	General Electric T-700 GE 401C	Range:	300 nautical miles
Take-off power:	2 × 1713 shp	Service ceiling:	5000 ft hover

SERIAL INFORMATION

'6001' to '6032' US Navy BuNo's 163801 to 163832.

Deliveries:

'6001' US Naval Air Test Center, Patuxent River, Maryland. Late 1989. Specification compliance testing until April 1991. USCG Air Training Center, Mobile, Alabama.

'6002' USCG Aircraft Repair & Supply Center, Elizabeth City, North Carolina. Aircraft Computerized Maintenance System (ACMS). Delivered June 1990. Late August 1990 to AEL Corporation, Allaire, New Jersey for improved HF antenna installation, followed by testing at NATC Patuxent River. Aviation Technical Training Center (ATTC) USCG Elizabeth City, North Carolina.

'6003' August 1990 to USCG ATTC., Elizabeth City.

'6004' September 1990. USCG Air Training Center, Mobile, Alabama.

Deployment

US Coast Guard Air Station,	Elizabeth City, NC.	June 1991
	Traverse City, Michigan	September 1991
	San Francisco, California	January 1992
	Clearwater, Florida	March 1992
	Cape Cod, Massachusetts	April 1994
	Sitka, Alaska	November 1994
	Kodiak, Alaska	July 1995

STEARMAN N2S-3

Commencing as a private venture in 1934, this famous series of biplane trainers, which perhaps could claim to be the world's most produced biplane, entered US Coast Guard service during World War II, eleven being used as proficiency training aircraft and also for courier duties. It was well known for its aerobatic qualities. Although limited in range the records reveal they were used very occasionally for search and rescue missions.

In 1934, when Stearman also became a Boeing subsidiary, the Model 70 was built as a potential replacement for the earlier Consolidated PT-1, PT-3 and PT-11 primary trainers in use by the US Army Air Corps. The new Stearman was a conventional biplane with a 225 hp Wright or a 215 Lycoming radial engine.

The Model 70 won an Army Primary Trainer contest during 1934, first production orders not coming from the Air Corps but the US Navy with proviso that the older Wright R-790-8 engine be installed in order to use up existing stocks. In this guise it became Model 73 carrying the US Navy Designation NS-1, with some sixty-one being delivered during 1935/36.

Improvements were incorporated in the Stearman Model 75 powered with a Lycoming engine which appeared in 1936, this being the first of the series ordered by the US Army Air Corps, and designated PT-13. Over the next ten years, well over 8000 Model 75s were built, the majority for the US Army Air Force and the US Navy, who eventually adopted a standardised model, which was fully interchangeable between the two services. The first US Navy order involved 250 N2S-1 trainers similar to the PT-17 and powered by the Continental R-670-14 engine. A further 125 powered by Lycoming R-680-8 engines were designated N2S-2 by the US Navy, becoming the Stearman Model B-75.

Eleven Stearman N2S-3 biplane trainers were used by the USCG during World War II, remaining on the inventory until 1947. It was known in the US Army Air Force as the PT-17. Depicted is a N2S-3 trainer of the US Navy, no photo of a USCG aircraft being available. *(William T Larkins)*

A further engine change, to the Continental R-670-4 identified the N2S-3 — Model A75N-1 — of which the US Navy acquired 1875. Eleven N2S-3 trainers acquired by the Coast Guard during 1943 were retained on the USCG aircraft inventory until 1947. It is one of the few Coast Guard aircraft for which no serial numbers are available, no photographs and no details of their use within the service.

TECHNICAL DATA

Manufacturer:	Stearman	Take-off power:	1 × 220 hp
Other designations:	PT-17	Designation:	N2S-3 Kaydet
Span:	32 ft 2 in	Type:	Trainer
Height:	9 ft 2 in	Length:	25 ft
Fuel:	43 galls	Wing area:	297 sq ft
Empty weight:	1936 lb	Oil:	4 galls
Crew:	2	Gross weight:	2717 lb
Top speed:	124 mph at sea level	Stall speed:	50 mph
Cruise:	106 mph at sea level	Range:	505 miles
Sea-level climb:	840 ft per min	Service ceiling:	11,200 ft
Engine:	Continental R-670-4		

STINSON RQ-1 XR3Q-1 RELIANT

Two four-seat Stinson SR-5 Reliant cabin monoplanes were acquired by the US military during 1935, one for the US Navy, the other for the US Coast Guard. At this time the US Navy had taken on charge a number of commercial aircraft for use on light transport liaison duties.

Only one Stinson Reliant served with the USCG, being commissioned during 1935, serving until 1941. The photo shows RQ-1 '381' wrongly designated QR-1 in USCG livery and 'RADIO TEST' on the entrance door confirming its role whilst based at Floyd Bennett Air Station, New York. *(William T Larkins)*

The Reliant for the Coast Guard was initially designated RQ-1, being later redesignated XR3Q-1. It was registered 381 by the Coast Guard, later becoming V149 and was commissioned on 20 February 1935, being based at Floyd Bennett Field, Brooklyn, New York. It was employed on electronic equipment test flying. It had a cruise speed of 120 knots, and had the Stinson c/n 74. It was decommissioned in 1941. During 1939 it was based with the Air Patrol detachement located at Cape May, New Jersey.

TECHNICAL DATA

Manufacturer:	Stinson
Other designations:	SR-5 Reliant
Contract No.	Tcg. 23248
Span:	41 ft
Height:	8 ft 8 in
Fuel:	50 gal
Empty weight:	2250 lb
Crew:	4
Top speed:	133 mph
Cruise:	123 mph
Engine:	Lycoming R-680-6
Take-off power:	1 × 225 hp
Prop diameter:	8 ft 6 in
Designation:	XR3Q-1
Type:	Electronic test
Unit cost:	$11,370
Length:	27 ft 3 in
Wing area:	230 sq ft
Gross weight:	3550 lb
Stall speed:	64 mph
Range:	350 miles
Service ceiling:	13,200 ft
Prop blades:	Smith, controllable pitch

STINSON OY-1/2 SENTINEL

The Stinson OY-1 Sentinel was used by the US Coast Guard on general duties, including spotting illicit stills for the US Alcohol Tax Unit. Four OY-1 aircraft were acquired during September 1948, plus three L-5 Sentinels for cannibalisation to provide spare parts. During May 1949, three of the four OY-1s were retired, and a single OY-2 was added to the inventory in 1952. The main base for both OY-1 and OY-2 aircraft was Elizabeth City air station in North Carolina, although it is known that at least one served on TDY (temporary duty) at San Diego air station in California during 1948.

One of the many World War II US Army light aircraft types produced and used on liaison and communication duties was the Stinson L-5 Sentinel. This was a development of the Stinson 105 Voyager, then in production during 1941 as a three-seat lightplane for the commercial market. The Army version which was procured during 1942, and designated O-62, had a larger fuselage, a greater all-up weight, but retained the 34ft wingspan. It was powered by a Lycoming 185hp engine. US Army trials had evaluated six commercial 105 Voyagers which they purchased in 1941 with the designation YO-54.

The L-5 Sentinel became the second most-used of all the US

Four OY-1 Sentinel light liaison aircraft were acquired by the USCG and utilised on general duties, including spotting illicit stills. Depicted is OY-1 98168 which for a time was based at Elizabeth City, North Carolina. *(Howard Levy)*

Army liaison aircraft. Initial deliveries totalled 1731, and orders for later variants brought the total to well over 3000. After the first 275 had been delivered the designation was changed from 0-62 to L-5. During 1943, 688 of the L-5s were converted to L-5A standard with a 24-volt electrical system. The undercarriage leg fairings were removed from all L-5s.

An ambulance version designated L-5B was modified with an upward-hinged hatch aft of the cabin with provision for a stretcher. During World War II this version served in the Pacific theatre of operations, later in the Korean conflict. A total of 679 were built. Two hundred L-5C aircraft had a K-2 reconnaissance camera in the fuselage, and the 558 L-5E Sentinels were similar to the L-5C, but had drooping ailerons which operated in conjunction with the flaps. A single XL-5F had a 0-435-2 engine and other modifications. The final production version was the L-5G of which 115 were built, powered with a 190hp 0-435-11 engine.

The US Marine Corps acquired 306 Convair-built Stinson Model V-76 aircraft, designated OY-1. This was the L-5 as in service with the US Army, built after the Stinson company had become a division of Vultee Aircraft. These 306 were L-5B and L-5Es transferred as the OY-1 with new serials. In December 1948, all OY-1s with a 24-volt electrical system were redesignated OY-2. The OY-2 Sentinel was used to establish direct support of US Marine Corps observation squadrons and served in Korea until April 1952. Stinson continued as a division of Vultee until March 1943, when it became the Stinson Division of Convair — Consolidated Vultee. During World War II the US Marine Corps used the type on general liaison and light transport duties, plus artillery spotting and casualty evacuation in forward combat areas.

As mentioned earlier the duties of the US Coast Guard OY-1 aircraft included spotting illicit stills for the Alcoholic Tax Unit of the UN Internal Revenue Service. Four were used operationally while the remaining three were cannibalised for spares. One of the aircraft based at the Elizabeth City air station was 98168 ex-US Air Force L-5-VW 42-98168, while the OY-1 recorded at San Diego air station on October 25, 1948, was 14870 ex-US Air Force L-5-VW 42-14870.

TECHNICAL DATA

Manufacturer:	Stinson — Convair	Engine:	Lycoming 0-435-A
Other designations:	L-5	Take-off power:	1 × 185 hp
Span:	34 ft	Designation:	OY-1/2 Sentinel
Height:	8 ft 11 in	Type:	Observation
Fuel:	36 gal	Length:	24 ft 1 in
Empty weight:	1550 lb	Wing area:	155 sq ft
Crew:	2	Gross weight:	2185 lb
Top speed:	129 mph	Range:	420 st miles
Cruise:	115 mph	Service ceiling:	15,800 ft
Sea-level climb:	975 ft per min		

VOUGHT UO-1/4

By 1925, the US Coast Guard was engaged in attempting to stop a massive evasion of the Eighteenth Ammendment of the Constitution — the prohibition of the manufacture, sale and transportation of alcholic beverages. Rum-running became so flagrant that surface craft were hard pressed. Again, the Coast Guard demonstrated the usefulness of aviation, this time with official notice and action.

Early in 1925 Lieutenant Commander C G von Paulson, supported by the Coast Guard Commandant, Rear Admiral Frederick C Billard, obtained the loan of a Vought UO-1 seaplane from the US Navy for one year. For a time this seaplane was based at the US Naval Reserve Station, Squantam, Massachusetts. Then it was operated from a small base established on Ten Pound Island in Gloucester Harbor. A schedule of daily patrol flights substantially curtailed the rum-running in the area. As a sideline to the patrol flights, the staff at the base gave Coast Guard aviation students instruction, and performed a number of interesting experiments in the use of radio communications between aircraft in flight and between aircraft and ship-ground

The USCG operated two Chance-Vought UO-4 aircraft, both commissioned in December 1926, being modified UO-1 seaplanes for the Coast Guard. Initially they were registered 'CG4' and 'CG5'. Depicted in flight is one of the UO-4 seaplanes in early USCG livery. *(USCG)*

stations. One of the most important achievements in this latter area was the development of the first loop-type radio direction finder.

Impressed by the activity of the air station at Ten Pound Island and plagued by the increasing operations of rum-runners in other areas, the US Congress appropriated $152,000 for the purchase of five aircraft for the Coast Guard. These were the very first the USCG could claim as its own, all previous equipment having been borrowed from the US Navy.

During 1926, two modified Vought UO-1 seaplanes were built for the Coast Guard with the designation UO-4. These were seaplanes using the new 220hp Wright J-5C Whirlwind engine and were fitted with the improved UO-3/FU-1 wings. Delivered in December 1926, they were given the Coast Guard numbers 4 and 5, later changed to 404 and 405. Vought UO-4 404 was still in service during 1935. They later became registered V104 and V105.

TECHNICAL DATA

Manufacturer:	Vought	Prop diameter:	8 ft 9 in twin-blade
Other designations:	Spec. SD-73-3	Designation:	UO-4
Span:	34 ft 4 in	Type:	Observation biplane
Height:	10 ft	Unit cost:	$18,000
Fuel:	46 gal	Length:	28 ft 5 in
Empty weight:	1860 lb	Wing area:	290 sq ft
Crew:	2	Oil:	4½ galls
Top speed:	122 mph	Gross weight:	2779 lb
Cruise:	90 mph	Stall speed:	61 mph
Sea-level climb:	930 ft per min	Range:	365 miles
Engine:	Wright J-5C Whirlwind R-790	Service ceiling:	14,900 ft
		Prop blades:	Steel-fixed pitch
Take-off power:	1 × 225 hp		

SERIAL INFORMATION

USCG No	Commissioned	Decommissioned
104 V104	Dec 1926	UO-4 ex-5 later 404
105 V105	Dec 1926	UO-4 ex-5 later 405

VOUGHT O2U-2 CORSAIR

Combining practical knowledge gained with earlier Vought models, of which the US Coast Guard already had experience, the UO-1 and UO-4, the Chance Vought Corporation designed a new observation biplane during 1926, which was powered by the new Pratt & Whitney Wasp engine. It was the first US Navy aircraft designed around this famous power plant, and one of the first aircraft to have all-steel-tube fuselage construction. Features which were retained from earlier designs included the cheek tank and the method of streamlining the fuselage. Two prototypes were ordered by the US Navy during 1926, designated O2U-1, to be followed by 130 production aircraft, also designated O2U-1, which commenced delivery to the US Navy during 1927. The famous name Corsair was adopted.

The O2U-1 carried a crew of two in open tandem cockpits, with provision for one fixed forward-firing gun and a Scarff ring for one or two machine guns in the rear cockpit. There was provision for bomb racks under the wings. In addition to operating from land bases or aircraft carriers with its wheel chasis, the Corsair was designed for use from battleships and cruisers as a floatplane with a single float and wingtip stabilisers. During 1928 an amphibious float was developed for the O2U-1, consisting of wheels on each side of the main float which retracted upwards to house partially submerged in the top of the float when operated as a floatplane.

This O2U-2 Corsair '303' possibly c/n 8109, was one of four delivered to the USCG during 1934/35, and was later registered 'V119' surviving until April 1937. They were observation aircraft and served with the air patrol detachment in Texas. *(USCG)*

By mid-1928 the O2U-1 Corsair became standard equipment with the US Navy observation squadrons which provided aircraft for the battleship divisions. They also equipped US Marine Corps squadrons. Production continued into 1928 with the O2U-2 model of which thirty-seven were built and was distinguished by minor alterations which included dihedral on the lower wing, a different cut-out shape in the upper centre section and a larger rudder.

A total of six Vought O2U-2 Corsairs were delivered to the US Coast Guard, three during 1934 and three the following year these being land biplanes. They were purchased by the US Navy on behalf of the Coast Guard and allocated the serial numbers '301 to 306' later being changed to V117 to V122. Several were stationed at San Antonio, Texas, with the USCG Air Patrol for the prevention of illegal immigration.

Four of the six Corsairs were disposed of or surveyed, to quote USCG paperwork, during 1937, but V118 survived until January 1940.

TECHNICAL DATA

Manufacturer:	Vought	Type:	Observation biplane
Contract No.	Via US Navy	Length:	24 ft 6 in
Span:	34 ft 6 in	Wing area:	319 sq ft
Height:	11 ft	Oil:	8 galls
Fuel:	110 gal	Gross weight:	3703 lb
Empty weight:	2252 lb	Stall speed:	53 mph
Crew:	2	Range:	450 miles
Top speed:	147 mph	Service ceiling:	20,100 ft
Cruise:	90 mph	Armament:	One fixed 0·30 in machine gun Two flexible 0·30 in on Scarf ring.
Sea-level climb:	1310 ft per min		
Engine:	Pratt & Whitney Wasp R-1340-88		
Take-off power:	1 × 450 hp	Prop blades:	Standard steel, 3792, fixed pitch
Prop diameter:	8 ft 11½ in twin-blade		
Designation:	O2U-2		

SERIAL INFORMATION

USCG No	Commissioned	Decommissioned
301 V117	Jul 1934	Aug 1934
302 V118	Aug 1934	Jan 1940
303 V119	May 1934	Apr 1937 Surveyed
304 V120	Feb 1935	Apr 1937 Surveyed
305 V121	May 1935	Jun 1937 Surveyed
306 V122	Mar 1935	Jun 1937 Surveyed

VOUGHT OS2U-3 KINGFISHER

During World War II the ceaseless guarding of vital convoys by aircraft and combat cutters of the US Coast Guard contributed greatly to a reduction of ship losses. The Vought VS-310 Kingfisher, one of several observation and scouting aircraft available to the US Navy when the USA entered World War II, proved to be the most useful aircraft used by both the US Navy and the Coast Guard.

This versatile observation-scout aircraft constructed by the Vought-Sikorsky Division of the United Aircraft Corporation at Stratford, Connecticut, incorporated some revolutionary structural techniques, including spot welding. The OS2U Kingfisher was based upon the Vought company's considerable experience of observation aircraft and was designed to replace earlier Vought types in a similar role. Layout included the use of a large single float plus underwing stabilising floats. The fuselage layout was similar to that employed in earlier biplane scouts. A tailwheel undercarriage could be fitted.

The contract with Vought to construct a prototype XOS2U-1 was placed by the US Navy on 22 March 1937, this appearing in 1938 powered by a 450hp Pratt & Whitney R-985-4 engine, its first flight taking place on 20 July 1938. By August 1940, the first of fifty-four OS2U-1 examples had reached the US fleet, with small changes to the float attachments and an R-985-48 engine. When the OS2U-2 model appeared during 1940/41 there were further small changes of equipment plus an R-985-50 engine. NAS Pensacola and Jacksonville, Florida, received the Kingfisher, while nine inshore patrol squadrons formed in 1942 received exclusively the OS2N-1 Kingfisher, these being similar to the OS2U-3 but built by the Naval Aircraft Factory located at Philadelphia, Pennsylvania.

These differed from earlier models in having extra fuel tanks in the wings and improved armour protection for the pilot and observer. Commencing in 1941, Vought delivered 1006 OS2U-3s before production ended in 1942. Fifty-three OS2U-3 Kingfishers were flown by Coast Guard pilots during World War II, mostly on coastal anti-submarine patrols and included a number of Naval Aircraft Factory built OS2N-1 aircraft. With one depth charge and a full load of fuel they could struggle to get airborne and remain on patrol for six hours. They carried the burden of the anti-submarine warfare efforts in costal waters. By 1944 all had been phased out of Coast Guard service.

OPERATIONS

On 3 April 1943, Lieutenant J N Schrader, while patrolling out of the Miami Coast Guard air station in a Vought OS2U-3 Kingfisher observation and scout aircraft, received a radio message to search for survivors of the torpedoed tanker

Gulfstate. Upon sighting the remains of the tanker, he was able to spot three groups of survivors. Dropping his depth charge, he landed in the ocean and picked up the three men in the first group. He taxied to the second group, gave them a rubber raft for support, and carried on to the assistance of the third group, one of who was badly injured with burns. Taking this man on board his already overloaded OS2U-3 aircraft, Schrader stood by to protect the drifting survivors from sharks until other Coast Guard aircraft arrived to assist him.

During 1943 three OS2U-3 Kingfishers were airborne from the air station at Elizabeth City, North Carolina, on a routine patrol when they spotted sailors in the water about thirty miles east of Cape Hatteras. All three aircraft made off-shore landings, and picked up all the survivors who rested on the wings until a boat arrived from Elizabeth City to pick them up. They were all survivors from a tanker torpedoed off Cape Hatteras.

The Vought OS2U-3 Kingfisher observation and scout aircraft was used in large numbers by the US Navy, with over fifty being operated by the USCG. They were used extensively on convoy patrol duties and served with the service between 1942/44 contributing greatly to the reduction in ship tonnage losses. *(USCG)*

TECHNICAL DATA

Manufacturer:	Vought	Length:	33 ft 10 in
Other designations:	OS2N-1	Wing area:	262 sq ft
Span:	35 ft 11 in	Oil:	10 galls
Height:	15 ft 1 in	Gross weight:	6000 lb
Fuel:	144 galls	Stall speed:	50 mph
Empty weight:	4123 lb	Range:	805 st miles
Crew:	2	Service ceiling:	13,000 ft
Top speed:	164 mph at 5500 ft	Armament:	One ·30 in gun fixed forward
Cruise:	119 mph at 5000 ft		One ·30 in gun flexible mounted
Engine:	Pratt & Whitney R-985-AN-2/8		Bombs — depth charges
Take-off power:	1 × 450 hp		
Designation:	OS2U-3 Kingfisher	Prop blades:	Hamilton-Standard constant speed
Type:	Observation-scout		

Vought OS2U-2/3 Kingfisher assignments with the US Coast Guard 28 February 1943.

Salem, Massachusetts BuNo.	5369, 5375, 5584, 5769, 5776, 5782, 5784, 5790.
Elizabeth City, North Carolina	5764, 5768, 5771, 5772, 5774, 5780, 5786, 5965.
Miami, Florida	5763, 5765, 5770, 5775, 5779, 5783, 5785, 5800.
St. Petersburg, Florida	5766, 5773, 5778, 5781, 5784, 5787, 5795, 5797, 5799.
Biloxi, Mississippi	5788, 5792, 5793, 5796, 5798, 5802.
San Francisco, California	2239, 2267, 2269, 2270, 3093, 3114.

The USCG assignment dated 30 November 1942, listed BuNo 5789 at Salem and BuNo 5801 at Miami.

Three OS2U-3 aircraft BuNo 5369, 5375 and 5584 were allocated to the USCG in exchange for three J2F-5 aircraft BuNo 00735, 00736 and 00739. OS2U-3 BuNo 09401 was delivered to the US Navy on 5 May 1942, and during August 1944, was based with USCG at Biloxi, Mississippi.

VULTEE SNV-1 VALIANT

World War II produced a series of low-wing trainers from the Nashville, Tennessee, factory of Vultee between 1940 and 1944 which outnumbered all other aircraft in the basic trainer type. The Vultee Model 54 was a private venture described by one aviation historian as a docile low-wing monoplane.

The US Army Air Corps selected the Vultee Model 54 during September 1939, which became the famed BT-13 Valiant. Built in one version the BC-3, with a retractable undercarriage and powered by a 600hp R-1340-45 engine followed. The Valiant was similar but had a 450hp R-985-25 engine and a fixed undercarriage. The US Army Air Corps ordered 300, which at the time was the largest placed for basic trainers, and was possibly one of the largest single orders for any aircraft type. However, during 1941, it was followed by two larger contracts of which one was for 2000 aircraft. These contracts were for the BT-13A, powered by a R-985-AN-1 engine and having small refinements. Eventually, production of this version totalled 6,407 for the US Army Air Forces. Also procured were no less than 1125 BT-13B aircraft fitted with a 24-volt system.

Engine production could not keep pace with aircraft production, so consequently 1693 Valiants were fitted with the 450hp Wright R-975-11 engine during 1941/42. this version became designated BT-15A. A BT-13A rebuild by Vidal during 1942 having an all-plastic fuselage became designated XBT-16. By 1945, the role of the basic trainer in the huge training programme was taken over by more advanced types, like the North American AT-6.

Vultee SNV-1 BuNo 12821 of the US Navy, similar to the two SNV-1 Valiants procured by the Coast Guard in 1942. This basic trainer was in full use by the US Army Air Corps and the US Navy. *(Peter M Bowers)*

An order dated 28 August 1940, procured 1350 SNV-1 Valiants for the US Navy. These Vultee Model 74s had Pratt & Whitney R-985-AN-1 engines and were equivalent to the US Army Air Corps BT-13A. During 1942 the Coast Guard procured two SNV-1 aircraft which became registered V222 and V223 and were purchased for $75,413 each. They were used for instrument flying training plus utility duties. They cruised at 145 knots and had a stall speed of 45 knots. They remained in service with the Coast Guard until 1945. Unfortunately the archives do not reveal details of the base used by the aircraft, their US Navy identities or their eventual fate. However it is known that both SNV-1s were assigned to Biloxi, Mississippi, and listed on the inventory of Coast Guard aircraft dated 28 February 1943.

TECHNICAL DATA

Manufacturer:	Vultee	Prop diameter:	9 ft Two-blade
Other designations:	BT-13A	Designation:	SNV-1 Valiant
Span:	42 ft	Type:	Basic Trainer
Height:	9 ft 1 in	Unit cost:	$75,413
Fuel:	92 gal	Length:	28 ft 7 in
Empty weight:	2976 lb	Wing area:	238 sq ft
Crew:	2	Oil:	10 galls
Top speed:	182 mph	Gross weight:	3991 lb
Cruise:	158 mph at 5000 ft	Stall speed:	53 mph
Sea-level climb:	1675 ft per min	Range:	1030 st miles
Engine:	Pratt & Whitney Wasp Jr. R-985-AN-1	Service ceiling:	21,000 ft
Take-off power:	1 × 450 hp	Prop blades:	Hamilton Standard — two position

WACO J2W-1

Since 1926 the Waco Aircraft Company of Troy, Ohio, was one of the leading US manufacturers of open-cockpit training and sports biplanes. The name Waco is an acronym resulting from the initials of a previous company, the Weaver Aircraft Company. During 1934 the US Navy took delivery of two Waco UBF three-seat open-cockpit biplanes modified as XJW-1s, being utility aircraft for the airship USS *Macon*. They were fitted with trapeze hooks for engaging on the airship, and were also used as training aircraft for checking out new US Navy pilots in hook-on procedures.

During March 1937 the US Coast Guard procured three 1936 cabin model EQC-6 biplanes designated J2W-1. They were versatile in that they could be used as a landplane, fitted with floats and with skis. Aircraft registered V159 was based on board the USCG Cutter *Spencer* during 1937 whose home port was Cordova, Alaska. As a landplane this same aircraft was based at Coast Guard air station Floyd Bennett Field, Brooklyn, New York. They were used for various assignments, all three eventually being attached to the Air Patrol Detachment based at El Paso, Texas. Unfortunately all three were destroyed in accidents during 1939.

During March 1937, the USCG received three Waco J2W-1 observation aircraft. They were very versatile and could be fitted with wheels, skis, or floats. Depicted is J2W-1 'V159' parked at the USCG Air Station located at Floyd Bennett Field, Brooklyn, New York. All were destroyed in crashes during 1939. *(USCG)*

TECHNICAL DATA

Manufacturer:	Waco	Designation:	J2W-1
Other designations:	EQC-6	Type:	Observation
Contract No.	Tcg. 26677	Unit cost:	$12,054
Span:	35 ft	Length:	25 ft 9 in
Height:	8 ft 8 in	Wing area:	244 sq ft
Fuel:	70 gal	Oil:	5 galls
Empty weight:	2050 lb	Gross weight:	3350 lb
Crew:	2	Passengers:	3
Top speed:	159 mph	Stall speed:	55 mph
Cruise:	140 mph	Range:	600 st miles
Sea-level climb:	760 ft per min	Service ceiling:	16,000 ft
Engine:	Jacobs L-4 R-755	Prop blades:	Hamilton Standard — two position
Take-off power:	1 × 225 hp		
Prop diameter:	8 ft 4 in Two-blade		

SERIAL INFORMATION

USCG No	Commissioned	Decommissioned
V157	Mar 1937	Jan 1939 Crashed
V158	Mar 1937	Apr 1939 Crashed
V159	Mar 1937	Oct 1939 Crashed